Monte Carlo N-Particle Simulations for Nuclear Detection and Safeguards

John S. Hendricks • Martyn T. Swinhoe •
Andrea Favalli

Monte Carlo N-Particle Simulations for Nuclear Detection and Safeguards

An Examples-Based Guide for Students and Practitioners

 Springer

OPEN ACCESS

John S. Hendricks
Los Alamos National Laboratory
Los Alamos, NM, USA

Martyn T. Swinhoe
Los Alamos National Laboratory
Los Alamos, NM, USA

Andrea Favalli
Los Alamos National Laboratory
Los Alamos, NM, USA

This book is an open access publication.

ISBN 978-3-031-04131-0   ISBN 978-3-031-04129-7   (eBook)
https://doi.org/10.1007/978-3-031-04129-7

This Springer imprint is published by the registered company Springer Nature Switzerland AG
The registered company address is: Gewerbestrasse 11, 6330 Cham, Switzerland

Preface

This book is intended as a tool for Monte Carlo N-Particle (MCNP®) software practitioners who use the code to simulate the interaction of radiation with matter in general and safeguards-related detector instrumentation in particular. Most MCNP software-related references are directed toward users who are modeling for other applications. Nuclear safeguards applications require a distinctly different approach to MCNP modeling. This book contains a fundamentals review, followed by examples written by safeguards modeling experts. It also includes an overview of advanced concepts like variance reduction. We hope that this guide will help new modelers learn to use the MCNP software for safeguards applications and save time and effort for the veteran safeguards modeler in working through the MCNP manual to solve problems that have already been solved.

This book does not introduce or explain safeguards concepts or techniques. The user is expected to be familiar with the type of instrumentation (both neutron and gamma) that is currently applied. For background information on nondestructive assay methods for safeguards, we refer the reader to other publications such as the "PANDA" manual[1] or the book on radiation detection by Knoll.[2]

This book is not intended to replace the MCNP software manual.[3] We will refer to the MCNP User's Manual Code Version 6.2 extensively. The examples in this book were run using MCNP version 6.2 software, but the general principles described have been applicable for many previous versions of the code and are

---

[1] "Passive Nondestructive Assay of Nuclear Materials" https://www.lanl.gov/org/ddste/aldgs/sst-training/technical-references.php

[2] G. F. Knoll "Radiation Detection and Measurement" John Wiley 2000.

[3] C. J. Werner, Ed., "MCNP User's Manual Code Version 6.2," LA-UR-17-29981 (October 27, 2017).

expected to be valid also in future versions. For general information on the code, its history and availability, as well as latest information and news, please see https:// mcnp.lanl.gov/.

Los Alamos, NM, USA   J. S. Hendricks

Los Alamos, NM, USA   M. T. Swinhoe

Los Alamos, NM, USA   A. Favalli

November 2021

Acknowledgments

The authors acknowledge the Office of International Nuclear Safeguards–Human Capital Development Subprogram for the support and the Los Alamos National Laboratory (LANL) Research Library for the support in publishing the book as an open-access book.

An important contribution to the quality of the manuscript was made by members of the XCP-3 group at Los Alamos National Laboratory, who reviewed the manuscript and made many constructive suggestions. Nevertheless, any errors and omissions remain the responsibility of the authors.

The authors would also like to thank Tamara Hawman, LANL Communications and External Affairs-Communication Arts and Services, for editorial and formatting support.

About This Book

The MCNP® software is a general-purpose Monte Carlo N-Particle code that can be used for transport simulations of neutrons, photons, electrons, ions, etc. In the field of nuclear safeguards and nonproliferation, MCNP software simulations are routinely used for many purposes such as the design and calibration of detector systems and the study of nuclear safeguards scenarios, including radiation emission properties of fresh fuel in shipping containers and spent fuel in storage pools. Current literature lacks a reference book or repository material that systematically explains MCNP software usage by example; thus, the purpose of this work is to augment current literature with a reference repository that teaches MCNP software usage through detailed examples, with a focus on nuclear safeguards applications. We explain both MCNP software input and output files and the technical details used in MCNP input file preparation, which are linked to the MCNP manual. Examples have been selected based on the real-world experience of the safeguards group at Los Alamos National Laboratory (LANL) and the feedback of students trained in MCNP software by LANL safeguards personnel.

We expect this text to be of great benefit to students, postdocs, and practitioners who work in the nuclear engineering and nuclear physics fields with an emphasis on the use of MCNP software in nuclear safeguards and nonproliferation. It is a means to transfer the knowledge that the authors have developed through decades of work with the MCNP software—in real safeguards scenarios—to a new generation of subject matter experts.

Supplementary files Each chapter comes with downloadable input files for the user to easily reproduce the examples in the text. The files corresponding to each chapter can be located via a link in the footnote on the first page of that chapter.

Trademarks MCNP® is a registered trademark owned by Triad National Security, LLC, manager and operator of Los Alamos National Laboratory. Any questions regarding licensing, proper use, and/or proper attribution of Triad National Security, LLC, marks should be directed to trademarks@lanl.gov.

Contents

About the Authors

John S. Hendricks has been one of the principal developers/leaders of MCNP development for more than 30 years. In addition, he has taught more than 100 MCNP classes and consulted internationally on effective use of MCNP. In 2007, he was elected fellow to the American Nuclear Society for his contributions to Monte Carlo development.

Martyn T. Swinhoe was awarded the Vincent J. DeVito Distinguished Service Award in 2017 from the Institute for Nuclear Materials Management for—among many other achievements—pioneering the use of MCNP software for instrument design and enhancing the analysis of existing datasets to draw additional conclusions for nuclear materials management.[1]

Andrea Favalli was elected fellow to the American Physical Society in 2020 for his outstanding application of the methods and underlying science of nuclear physics to the crucial issues of nuclear safeguards and security. His work has focused on nondestructive assay of nuclear materials, ranging from new analytical approaches to experimental work.[2]

---

[1] https://www.lanl.gov/science-innovation/science-highlights/2017/2017-07.php

[2] https://www.lanl.gov/discover/news-release-archive/2020/October/1014-aps-fellows.php

Chapter 1
Introduction

This book is intended to aid nuclear safeguards scientists and engineers in more effective use of the MCNP® radiation modeling computer code. It should also be of use to both beginning and advanced users of the MCNP software in other fields and to those interested in nuclear safeguards technology and computational modeling.

1.1 Nuclear Safeguards

Ever since the discovery of nuclear fission, it has been recognized that nuclear materials provide both tremendous benefit and tremendous risk to society. At first, nearly all nuclear activities were conducted by personnel who had security clearances in government installations under guard. The Atomic Energy Act of 1954 was amended to permit use of nuclear materials by private industry—with certain safeguards requirements. Since then, the use of nuclear materials for power production, medicine, food sterilization, space applications, environmental control, research, and more has grown rapidly. Concern has also grown about the misuse of nuclear materials for crude nuclear weapons and "dirty bombs," which spread dangerous radioactive materials.

In 1957, six nations signed a treaty that established the European Community for Atomic Energy, which later became Euratom [1]. The treaty established the first multinational safeguards system with the right to send inspectors with access at all times to all places, data, and any involved persons to any member state territory. These responsibilities were then given to the International Atomic Energy Agency (IAEA), which had been established by a statute of the United Nations in 1956 and came into force in 1957.

In 1968, the Nuclear Non-Proliferation Treaty (NPT) was opened for signatures and came into force in 1970. The treaty (1) obligates countries that do not have nuclear weapons to neither produce nor receive them; (2) obligates nuclear weapon countries not to help non-nuclear countries obtain nuclear weapons; (3) provides

© The Author(s) 2022
J. S. Hendricks et al., *Monte Carlo N-Particle Simulations for Nuclear Detection and Safeguards*, https://doi.org/10.1007/978-3-031-04129-7_1

assurance, through international safeguards, that nuclear materials are not diverted to making weapons; (4) facilitates the sharing of peaceful benefits of nuclear materials; and (5) declares that all parties are determined to seek further arms control and disarmament [2]. Since then, more than 190 countries have signed the treaty. Many "near-nuclear" nations conditioned ratification on development of satisfactory safeguards arrangements [3].

Los Alamos National Laboratory (LANL) has been a leader in nuclear safeguards since its founding in 1943 because of its nuclear weapon expertise and outstanding experimental and computational facilities. Today, nuclear nonproliferation is one of LANL's principal missions. LANL scientists and engineers work to safeguard nuclear materials by developing techniques and systems for nondestructive assay of nuclear and hazardous materials. Applications include nuclear materials control and accountability for domestic DOE nuclear facilities, nuclear waste disposal, material stabilization efforts, and international nonproliferation efforts [4]. Since 1980, every single IAEA inspector has been trained at least once at LANL [5].

1.2   Monte Carlo N-Particle Transport: MCNP Code

The Monte Carlo method finds solutions to mathematical problems using statistical sampling with random numbers. The idea of solving problems statistically is first recorded by Comte de Buffon in the mid-1700s. In the 1930s, Enrico Fermi first experimented with the Monte Carlo method while studying neutron diffusion. The modern Monte Carlo method was developed by John von Neumann, Stanislaw Ulam, Nicholas Metropolis, and Robert Richtmyer at LANL [6–9]. Monte Carlo methods were central to the simulations required for the Manhattan Project, though severely limited by the computational tools at the time [10]. The development of the Monte Carlo method is among LANL's top achievements [11].

The MCNP code is a direct descendent of those early efforts. In the 1950s and 1960s, the Los Alamos Monte Carlo codes were organized into a series of special-purpose codes, including MCS, MCN, MCP, and MCG. These codes were able to transport neutrons and photons for specialized LANL applications. In 1977, these separate codes were combined to create the first generalized Monte Carlo radiation particle transport software, the MCNP code, which also included electrons. In 1999, the first version of a variant, MCNPX, was released. MCNPX expanded the MCNP code from modeling neutrons, photons, and electrons to 34 sub-nuclear particles and light ions. In 2008, MCNPX was expanded to treat all 2205 heavy ions. In 2013, MCNPX was merged into the MCNP code (MCNP6.1). All calculations in this book were performed using MCNP6.2 code, which was released in around 2017.

Approximately 20,000 copies of the MCNP code have been distributed to users in government institutions, academia, and private industries worldwide [12].

The MCNP code models the interaction of radiation with matter. The particles are modeled with fully continuous energy in the range from 1 keV to 1 TeV/nucleon and from 0 MeV upward for neutrons. Everything is fully three-dimensional and time-dependent. Where nuclear data are available, these data libraries are used in full detail, enabling predictive answers. Where nuclear data are unavailable, various physics model algorithms available in the MCNP code are used. A wide variety of source and tally models are standard features in the code; it is only rarely necessary to modify the program to model specific radiation sources or detectors. There are also many variance-reduction methods available to rapidly converge to the correct statistical result.

Improvements in computers since the Manhattan project have made the Monte Carlo sampling millions of times faster. The MCNP code variance-reduction capabilities speed up convergence by millions of times more. Parallel computation enhances computational speed even further. Consequently, the Monte Carlo method is the preferred method to accurately model the interaction of radiation with matter. The MCNP code is used in a wide variety of applications—from nuclear power to homeland security, from outer space to deep underground well-logging, from microelectronics to the human body, defense, homeland security, and many other applications in addition to safeguards.

In the field of nuclear safeguards and nonproliferation, MCNP code simulations are routinely used for many purposes such as the design and calibration of detector systems and the study of nuclear safeguards scenarios, including radiation emission properties of fresh fuel in shipping containers and spent fuel in storage pools.

1.3 Objective

The objective of this book is to assist new and current users of the MCNP code to create and run simulations that are effective (in producing valid results) and efficient (in optimizing both the user's and computer's time) by presenting examples that are relevant to the field. The focus is on the field of nuclear nonproliferation safeguards. Chapter 2 begins with the basic concepts of MCNP code calculations. This will serve as an *aide memoire* for experienced users. Less experienced MCNP users may need to supplement this chapter with MCNP primers such as the MCNP manual [13] or Shultis and Faw [14]. Chapter 3 presents specific examples of simulations of nuclear instrumentation. Experienced users may wish to start here. Chapter 4 contains examples of advanced concepts such as variance reduction and detailed detector modeling. Chapter 5 outlines some guidance on potential modeling errors and goes into more detail about delayed neutron production and comparing table physics to model physics.

References

1. F. Spaak, *The Euratom Safeguards System*, WASH-1149, Office of Safeguards and Materials Management, USAEC: Safeguards Papers from ANS/AIF Winter Meeting, Washington, DC, November 10–15, 1968 (1968), pp. 8–11
2. H. Scoville Jr., *The Negotiation of Safeguards in the Non-Proliferation Treaty*, WASH-1149, Office of Safeguards and Materials Management, USAEC: Safeguards Papers from ANS/AIF Winter Meeting, Washington, DC, November 10–15, 1968 (1968), pp. 56–59
3. M.B. Kratzer, *A New Era for International Safeguards*, Transactions of the American Nuclear Society, Winter Meeting, Washington, DC, November 15–19, 1970, vol 11 (1970)
4. Los Alamos National Laboratory (n.d.). https://lanl.gov/orgs/n/n1
5. Congressional Record, Congress. Rec. **162**(43), S1569–S1571 (2016) Government Publishing Office. www.gpo.gov
6. J.M. Hammersley, D.C. Handscomb, *Monte Carlo Methods* (John Wiley & Sons, New York, NY, 1964)
7. N. Metropolis, The beginning of the Monte Carlo method. Los Alamos Sci. **15**(Special Issue), 125–130 (1987) https://library.lanl.gov/cgi-bin/getfile?00326866.pdf
8. R. Eckhardt, Stan Ulam, John von Neumann, and the Monte Carlo method. Los Alamos Sci. **15**(Special Issue), 131–137 (1987) https://permalink.lanl.gov/object/tr?what=info:lanl-repo/lareport/LA-UR-88-9068
9. N.G. Cooper, *From Cardinals to Chaos: Reflections on the Life and Legacy of Stanislaw Ulam* (Cambridge University Press, New York, NY, 1989)
10. Wikipedia (n.d.). https://en.wikipedia.org/wiki/Monte_Carlo_method
11. Los Alamos National Laboratory (n.d.). https://www.lanl.gov/about/history-innovation/innovation-timeline.php
12. A. Sood, *The Monte Carlo Method and MCNP – A Brief Review of Our 40 Year History*, Presentation to the International Topical Meeting on Industrial Radiation and Radioisotope Measurement Applications Conference, Chicago, IL (10 July 2017) (2017), pp. 1–22. https://mcnp.lanl.gov/pdf_files/la-ur-17-26533.pdf
13. C.J. Werner, *MCNP User's Manual Code Version 6.2*, LA-UR-17-29981 (2017)
14. J.K. Shultis, R.E. Faw, *An MCNP Primer*, Kansas State University. https://www.mne.k-state.edu/~jks/MCNPprmr.pdf

Chapter 2
Basic Concepts

In this section, we will explain the basic MCNP® concepts necessary to prepare the input deck for the MCNP code to solve radiation transport problems. Here we assume that the readers have the MCNP software and a text editor available on their computers.

2.1 Geometry

2.1.1 Simplest Possible Input File

The MCNP code [1–3] runs by reading an ASCII input file that describes the problem geometry, materials, sources, tallies, physics, and options. Example 2.1 is the MCNP input file—the simplest possible input file.

Example 2.1 The Simplest Possible Input File

```
Trivial Example
1 0 -11 imp:n=1
2 0 +11 imp:n=0

11 SPH 0 0 0 5

SDEF
```

Supplementary Information The online version contains supplementary material available at [https://doi.org/10.1007/978-3-031-04129-7_2].

J. S. Hendricks et al., *Monte Carlo N-Particle Simulations for Nuclear Detection and Safeguards*, https://doi.org/10.1007/978-3-031-04129-7_2

The MCNP input deck consists of card images. Each line is called a card for historical reasons.

The first card of an MCNP input deck is the title card. Here, the title is "Trivial Example." The title line is followed by three sections of cards: cells, surfaces, and data. These sections are separated by a blank card. In Example 2.1,

```
1 0 -11 imp:n=1
```

specifies cell 1 with material 0 (vacuum) inside surface 11 with neutron importance 1 (see section 1.3.4.2 of the MCNP manual).

```
2 0 +11 imp:n=0
```

Cell 2 has material 0 and is outside surface 11 with importance 0, which means any particle that passes into cell 2 is no longer followed ("killed"). Thus cell 2 is the outside world. All MCNP geometries extend out infinitely, so cell 1 is everything inside surface 11 and cell 2 is everything outside.

After the blank line delimiter, the surface card(s) are defined.

```
11 SPH 0 0 0 5
```

describes surface 11 as a sphere at coordinates (0,0,0) with radius 5 cm (MCNP manual 3.2.2).

After the blank line delimiter, the data cards are defined. MCNP has hundreds of data cards that are defaulted but, at minimum, the source definition, SDEF, must be provided. The defaults are 14 MeV neutrons isotropic at the origin (0,0,0).

The default units for length, energy, time, and mass are cm, MeV, shakes (1×10^{-8} s), and grams, respectively (MCNP manual 7.1.6).

Comment cards begin with a c followed by a space and may be anywhere after the title card. Each card may also have a comment at the end started by "$." The "$" and what follows are ignored. Capitalization is ignored except when upper or lower case letters appear in the name of a file. Anything entered after column 128 is ignored. A card beginning with five or more blank spaces is a continuation of the previous card. A card that ends with "&" is continued on the next line regardless of whether the five or more blank spaces are present on that line. Entries may appear in any column otherwise.

2.1.2 Running MCNP: The Simplest Case

The input file is run from a Windows (or from a shell in LINUX/UNIX systems) command prompt as:

```
MCNP6     i=EX11_1
```

where the name of the text input file in this case is EX11_1 (beware of hidden "txt" extensions in Windows systems). This problem will run until terminated by an interrupt. The possible interrupts are

`<ctrl-c><s>`	See status of calculation—how many histories are run, etc.
`<ctrl-c><q>`	Quit cleanly—finishing last history and writing output files
`<ctrl-c><m>`	Interrupt—to plot geometry or tallies
`<ctrl-c><k>`	Kill run—do not use unless problem is hung

The calculation will generate an output file, OUTP, and a continue-run file, RUNTPE. If the problem was terminated by $<ctrl-c><q>$, it can be continued by

```
MCNP6   c   runtpe=RUNTPE
```

The output file from the continued run would be OUTQ (see section 1.4.1 in the MCNP manual for file naming).

The problem and its output files may also be renamed. For example, the initial run could be

```
MCNP6   i=EX11_1   n=XYZ.
```

Then the output file would be XYZ.o and the runtpe would be XYZ.r; the continuation would be

```
MCNP6   c   runtpe=XYZ.r   outp=ABC.o
```

In this example, the runtpe XYZ.r would contain the information from the first run and add, not overwrite, the information from the continuation. The continue-run output would be ABC.o.

A number of messages are printed by MCNP to the command window.

```
comment.   Physics models disabled.

comment.  using  random  number  generator   1,  initial  seed  =
19073486328125

comment.  total nubar used if fissionable isotopes are present.
```

Comments generally state what defaults are being used and usually can be ignored.

```
warning.  there are no tallies in this problem.
warning.  no cross-section tables are called for in this problem.
```

Warnings indicate something that looks wrong but may be intentional.

Fatal errors are user errors found while processing the input file. After the first fatal error, subsequent ones may be spurious because they are caused by an earlier fatal error.

Bad trouble errors are issued while histories are being run and indicate imminent catastrophe. Rather than crash and lose all output, etc., the bad trouble error immediately stops the calculation and puts out what is available so far.

The OUTP output file of this simple problem does not contain much because the problem is so simple. After the copyright notice is

```
1mcnp     version 6      ld=06/09/17                  05/15/19 13:31:20
 ***************************                probid =  05/15/19 13:31:20
  i=EX11_1 n=xyz.
```

The version and load date (ld=06/09/17) uniquely identify the version of MCNP being run followed by the time of the run. The probid (Problem Identification) is the unique identifier of the run, which is the time the run was started. The probid will be attached to all subsequent files for identification with this initial run. Next is the echo of the MCNP command. Here the input file name was EX11_1.

Following is a reprint of the input file and various print tables, which can be turned on and off with a PRINT card in the input file. The input file of Example 2.1 had no PRINT card, so only the minimum printout is provided. The more general print contains information about geometry, surfaces, sources, tallies, physics, data sources, variance reduction, and more.

The problem summary is the balance (creation and loss) of physical particles (weight), energy, and statistical samples (tracks). Absent from this input is what usually follows—namely summaries of all physics and statistical information in every problem cell and tally. The tallies are the answers requested by the user.

2.1.3 Simple Input File

The MCNP input file of Example 2.2 shows how to set materials and additional capabilities of the MCNP code to model a simple nuclear fuel rod. This input file describes a 300 cm long UO_2 fuel rod of radius 0.4840 cm encased in zirconium cladding with outer radius 0.5590 cm and is illustrated in Fig. 2.1.

Example 2.2 A Simple Input File

```
Fuel Pin
c *************************** cells ***************************
11 313 -10.44   -71    u=0 imp:n=1 $ Fuel
12 201  -6.55   -72 71 u=0 imp:n=1 $ Cladding
13 0                72 u=0 imp:n=0 $ Outside World
```

```
c *************************** surfaces ************************
71 RCC 0 0 -150    0 0 300    .4840    $ fuel cylinder
72 RCC 0 0 -150    0 0 300    .5590    $ clad cylinder

c ************************** data ****************************
SDEF
Print
NPS 1e6
m201   40090 -0.505239    40091 -0.110180      40092 -0.168413
       40094 -0.170672    40096 -0.027496      nlib=80c
m313   92235.80c -0.02759    92238.80c -0.85391  8016.80c -0.11850
```

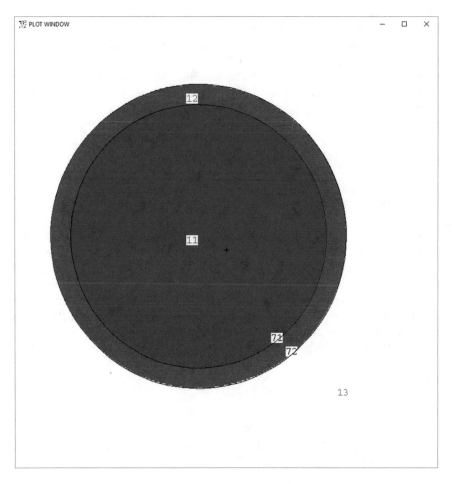

Fig. 2.1 $X - Y$ view of Example 2.2 geometry as plotted by the MCNP plotter. Labels 11–13 label the cell numbers and labels 71–72 the surface numbers, as specified in the EX11_2 input file

There are three comment cards in the input file of Example 2.2. The first line of Example 2.2 input file is the title card; the title is "Fuel Pin."

Cell cards other than comment cards have the following format: *cell number, cell material, cell density, surface relationships, and data*. Cell, surface, and material numbers are all arbitrary. Densities are positive for units of atom/(barn-cm) and negative for g/cm^3.

```
11 313 -10.44    -71     u=0 imp:n=1 $ Fuel
12 201  -6.55    -72 71 u=0 imp:n=1 $ Cladding
13 0                      72 u=0 imp:n=0 $ Outside World
```

Cell 11 has material 313, density 10.44 g/cm^3, inside surface 71, and neutron importance imp:n=1. (The u is explained below.) Cell 12 has material 201, density 6.55 g/cm^3, inside surface 72, and outside surface 71, with importance 1. Cell 13 has material 0, which is a void, in which case no density is specified. Cell 13 is outside surface 72 with importance 0. Note that the space outside surface 72 extends forever; all of space out to infinity must be specified. But because the importance is imp:n=0, any neutron that crosses surface 72 into cell 13 escapes the problem and is killed. Here, u=0 is a placeholder; it will be used to build fuel pin assemblies (see Sect. 2.1.7).

Surface cards start with a surface number, followed by a mnemonic surface type, and then the numerical entries for that surface type.

```
71 RCC 0 0 -150    0 0 300    .4840     $ fuel cylinder
72 RCC 0 0 -150    0 0 300    .5590     $ clad cylinder
```

Surface 71 is a right circular cylinder, RCC. The base is at $x, y, z = 0, 0, -150$ cm. The height vector is 0, 0, 300, corresponding to axis direction cosines $u, v, w = 0, 0, 1$, namely the z-axis with height 300 cm. The radius of surface 71 is 0.4840 cm and the radius of surface 72 is 0.5590 cm.

Thus, the problem geometry is two concentric cylinders with radii 0.4840 cm and 0.5590 cm. These form three cells: the fuel inside surface 71, the clad between surfaces 71 and 72, and the rest of the world outside surface 72.

After the cell and surface cards are the data cards:

```
SDEF
Print
NPS 1e6
m201    40090 -0.505239    40091 -0.110180    40092 -0.168413
        40094 -0.170672    40096 -0.027496    nlib=80c
m313    92235.80c -0.02759 92238.80c -0.85391 8016.80c -0.11850
```

Data cards may appear in any order. SDEF is the default source definition. Print provides a full rather than partial print in the OUTP output file. NPS runs the

problem for one million (1e6) source histories unless the problem is otherwise inter-
rupted. The number, NPS, of source histories, often referred to as "source particles"
by MCNP, is the number of times the problem source distribution will be sampled.
The number of histories is statistical and has no physical meaning. Note that by
default, SDEF has WGT = 1. *The particle weight (WGT) represents the physical
number of particles modeled per source particle and has no statistical meaning.*
Thus NPS 1e6 WGT = 1.0 samples the behavior of 1.0 physical source particle a
million times; NPS 1000 WGT = 6.02E23 samples the behavior of 6.02E23 physical source particles a thousand times.

Material 201 is specified next and is continued on the following line. Material
313 follows. Materials and cross sections are described in Sect. 2.2.

2.1.4   Running and Plotting MCNP Geometries

Input file EX11_2 of Example 2.2 can be run as

```
MCNP6    i=EX11_2 n=problem02.  IPXRZ
```

This problem will generate output file problem02.o and continue-run file
problem02.r, which are more readily associated with input file EX11_2 than the
default OUTP and RUNTPE names. It is recommended that the name option, name=
or n=, is used to give more meaningful file names.

IPXRZ is the execution line option: process input I, plot geometry P, get the
cross-section data X, run the problem R, and then plot the tallies Z. Usually it is
impractical to use all five of these options together. IXR is the default. IP can be
used to simply plot the geometry. The X–Y view MCNP plot of Example 2.2 geometry is shown in Fig. 2.1.

To get the plot in Fig. 2.1, run MCNP with

```
MCNP6    i=EX11_2 n=plot01.  IP
```

When the plot screen pops up, click the following buttons:

XY	Will get X–Y view
L2 Off	Will plot cell labels
10	Upper right corner—will zoom in by a factor of 10
10	Upper right corner—will zoom in by another factor of 10

Alternatively, when the plot screen pops up, click Plot> bottom center, and then
enter the following commands after the plot> prompt that appears in the same window where the MCNP code was run:

```
PZ=0   EX=1    LA=2 2 CEL
```

Note that the cell labels are twice as large because of the LA 2 2 CEL command; entering plot commands rather than clicking on the plot enables more capabilities. While in the plot> mode, the commands HELP, ?, and OPTIONS provide a help package that lists all possible geometry plot commands. To return to the interactive mode, enter INTERACT. To exit, click END in the interactive mode; type END in the plot> mode.

Plot> commands, such as EX=1, may also be entered at the lower left of the screen where it says "Click here or picture or menu." If this box is accidentally clicked, then everything is frozen until a command is entered there.

The other commands in the lower left are:

CURSOR	Select a part of the picture in which to zoom in
RESTORE	Return to previous picture
CellLine	Outline cells or show various tally and weight window meshes
PostScript	Make a PostScript format file of the picture
ROTATE	Rotate picture 90 degrees counterclockwise; same as "Theta 90" Plot> command
COLOR	Turn color on or off; click a cell quantity on the right margin, such as DEN, and colors will be by density rather than material
SCALES	Geometric scale in cm: 0 = none, 1 = on boundary; 2 = everywhere (grid)
LEVEL	Repeated structures universe level to view
XY, YZ, ZX	Plot coordinate axis view
L1 sur	On/off surface labels
L2 cel	On/off cell labels. Click on right to label by cel, mat, den, etc.
MBODY	On/off to plot macrobody facet numbers
LEGEND	Numerical values of shaded colors

The region in the upper left is not interactive, but if the screen goes blank this region can be clicked to restore the previous picture.

The controls at the top shift the screen up, down, right, or left. "Origin" moves the center of the plot screen to whichever part of the geometry is clicked. The scale from 0.1 to 10, top right, is a logarithmic scale zoom factor. Click it twice to zoom by that factor. Click it once and then click somewhere in the picture to zoom in or out of the clicked point in the picture.

Plotting is described in more detail in Sect. 2.5.1 and also in the MCNP6.2 code manual under the heading "5.2 THE GEOMETRY PLOTTER, PLOT."

The output file, problem01.o, like Example 2.1, begins with copyright, code version, date/time, probid, execution line, and input file echo. But because it is no longer a void problem but one with neutron transport physics, and because of the PRINT card, there is much more. The source, materials, volume, masses, surface areas, and surface descriptions are all listed. The cross sections for each material are described in Print Table 100 which appears in every output file if cross sections are used. *It is imperative to review Print Table 100 to ensure that the cross sections being used are the ones desired—in some circumstances, the MCNP code will select others!*

Print Table 110 lists the first 50 source particles, which helps determine if the source is set up as desired. The problem summary shows the gains and losses of the 1,000,000 statistical tracks and the physics of the average source particle. Additional tables show fission multiplicity, activity in each cell, and activity in each material.

2.1.5   Surfaces and Complicated Cells: Intersections and Unions

Example 2.3 and Example 2.4 describe a 12 cm high radius 5 cylinder inside a 1 cm thick box illustrated in Figs. 2.2 and 2.3. A million statistical histories are run with an isotropic 1 MeV source at $x, y, z = 0, 6, 0$ cm.

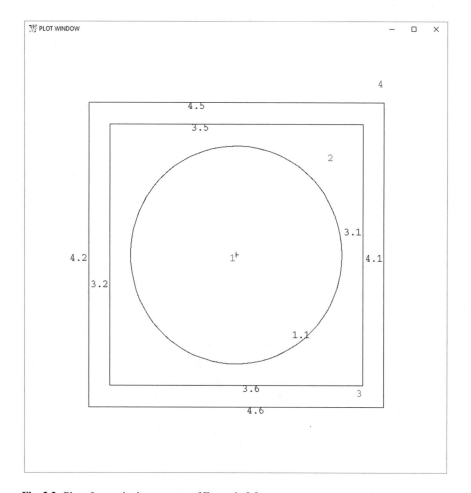

Fig. 2.2 Plot of macrobody geometry of Example 2.3

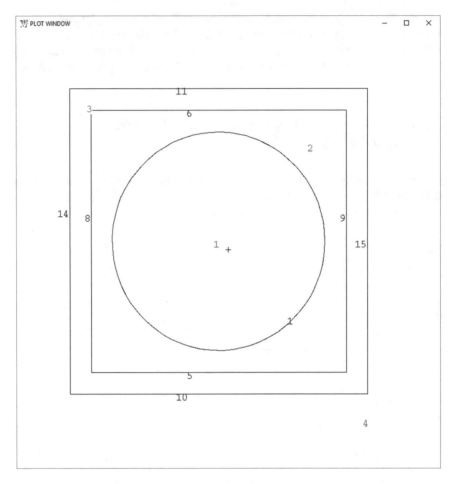

Fig. 2.3 Plot of quadratic surface geometry of Example 2.4

Example 2.3 Macrobody Description of Cylinder in a 1 cm Thick Box

```
Cylinder inside 1-cm thick box
1    0    -1           imp:n = 1  $ cylinder
2    0    1 -3         imp:n = 1  $ inner box
3    0    3  -4        imp:n = 1  $ outer box
4    0    4            imp:n = 0  $ Outside world

1 rcc   0 0 0 0 12 0 5  $ Right circular cylinder, 12 high, radius 5
3 rpp -6 6 0 14 -6 6  $ Right parallelepiped -6<x<6 0<y<14 -6<z<6
4 rpp -7 7 -1 15 -7 7   $ Right parallelepiped -7<x<7 -1<y<15 -7<z<7

sdef   pos   0 6 0 erg = 1
nps   1000000
```

Example 2.4 Quadratic Surface Description of Cylinder in a 1 cm Thick Box

```
Cylinder inside 1-cm thick box
1    0      -1 2 -3        imp:n = 1  $ cylinder
2    0 (1:-2:3)   5 -6 2 -7 8 -9    imp:n = 1  $ inner box
3    0 (-5:6:-2:7:-8:9) 10 -11 12 -13 14 -15 imp:n = 1 $ outer box
4    0   -10:11:-12:13:-14:15        imp:n = 0  $ Outside world

1    cy   5.0
2    py   0
3    py   12
c       inner box
5    px   -6
6    px    6
7    py   14
8    pz   -6
9    pz    6
c       outer box
10     px   -7
11     px    7
12     py   -1
13     py   15
14     pz   -7
15     pz    7

sdef  pos  0 6 0 erg = 1
nps   1000000
```

MCNP surfaces may be described by either macrobodies or quadratic surface equations. In a cell description, blanks between the surface numbers are the intersection operator, "AND," whereas the union operator, ":", between the surface numbers is "OR." Cells using the union operator are "complicated cells." Whereas macrobodies are internally converted to quadratic surfaces, cells using macrobody surfaces are also complicated cells.

The quadratic surface equations are described in Table 3-4 of the MCNP manual [2, 3]. They consist of planes P, PX, PY, PZ (arbitrary, normal to X, normal to Y, and normal to Z axis); spheres SO, S, SX, SY, SZ (centered at origin, anywhere, or along X, Y, or Z axis); cylinders C/X, C/Y, C/Z, CX, CY, CZ (parallel or on X, Y, Z axis); cones K/X, K/Y, K/Z, KX, KY, KZ (parallel or on X, Y, or Z axis); special quadratic SQ (ellipsoid, hyperboloid, paraboloid parallel to X, Y, or Z axis); general quadratic (cylinder, cone, ellipsoid, hyperboloid, paraboloid with arbitrary axis); and tori TX, TY, TZ (elliptical or circular torus on X, Y, or Z axis).

Surfaces may also be defined by coordinate points. Quadratic surfaces rotationally symmetric about the major axes are specified by the X, Y, and Z surface cards in (x,R) pairs, where x is the x, y, or z coordinate and R is the radius. One pair makes a PX,

PY, or PZ plane; two pairs make a CX, CY, CZ cylinder or KX, KY, KZ cone; three pairs make an ellipsoid, paraboloid, or hyperboloid. Skew planes may be specified on the general plane P surface card by providing three triplets of x, y, z coordinates.

It is easier and more common to describe MCNP surfaces by macrobodies described in Section 3.2.2.4 of the MCNP manual. These are BOX (planes specified by the three vectors at a corner); rectangular parallelepiped RPP (box of planes normal to X, Y, Z axes); sphere SPH (specifying X, Y, Z center and radius); right circular cylinder RCC (specifying X, Y, Z base and h1, h2, h3 height vector, and radius); and right hexagonal prism RHP, right elliptical cylinder REC, truncated right cone TRC, ellipsoid ELL, wedge WED, and arbitrary polyhedron ARB. The macrobodies are internally converted by the MCNP code into quadratic surface facets. Figure 2.2 shows the facets of macrobody surfaces 1, 3, and 4. The MCNP geometry plotter shows these by setting the MBODY option off.

The MCNP cell description uses senses to indicate whether a cell is inside or outside a surface. In the quadratic surface description of Example 2.4:

```
1    0    -1 2 -3  imp:n = 1  $ cylinder
```

Cell 1 with material zero (vacuum) is inside y-axis cylinder surface 1 cy 5.0, outside (above) y-axis plane surface 2 py 0, and below y-axis plane surface 3 py 12.

```
2 0 (1:-2:3) 5 -6 2 -7 8 -9  imp:n = 1 $ inner box
```

Cell 2 uses the union operator ":" to describe everything outside 1 or below 2 or above 3 (1:−2:3) combined with what is above 5 and below 6 and above 2 and below 7 and above 8 and below 9.

```
3 0 (-5:6:-2:7:-8:9) 10 -11 12 -13 14 -15
     imp:n = 1  $ outer box
```

Similarly, cell 3 is everything below 5 or above 6 or below 2 or above 7 or below 8 or above 9, combined with what is above 10 and below 11 and above 12 and below 13 and above 14 and below 15.

```
4    0    -10:11:-12:13:-14:15     imp:n = 0  $ Outside world
```

Cell 4 is the outside world extending to infinity and is everything below 10 or above 11 or below 12 or above 13 or below 14 or above 15.

The sense of a surface to a cell can be determined by picking any point in the cell, x, y, z and then using that point to evaluate the sense to the surface. For example, the z-axis CZ cylinder has the equation $x^2 + y^2 - R^2 = 0$; for radius $R = 10$ and x, y, $z = 1,2,3$, then $1^2 + 2^2 - 100 = -95 < 0$, so the cell containing this point has a negative sense relative to the surface; for $R = 10$ and x, y, $z = 8, 9, 0$, then $8^2 + 9^2 - 100 = 45 > 0$, so the cell containing this point has a positive sense relative to the surface. (Surface equations are given in the MCNP manual 3.2.2.1.)

2.1.6   Duplicate Cells, Complements, and Translations: LIKE n BUT and TRCL

Example 2.5 and Example 2.6 take the geometry of Example 2.3 and Example 2.4 and duplicate it in a 30° rotation translated 20 cm away in the X-direction. The source is now a 1 MeV isotropic point emitting at x, y, $z = 0$, 6, 0 and x, y, $z = 20$, 6, 0. Cells 11, 12, and 13 are the translated and rotated cells. See Figs. 2.4 and 2.5.

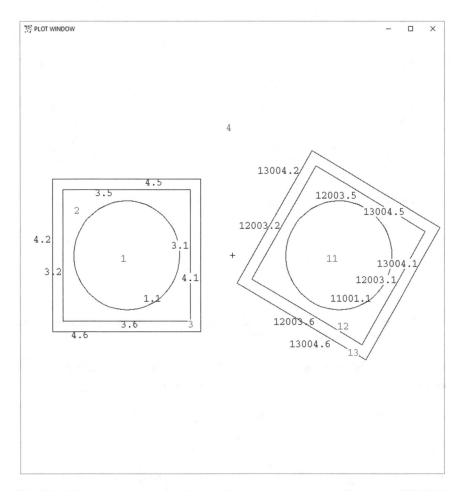

Fig. 2.4 Cylinder inside a 1 cm thick box duplicated into a second rotated box using LIKE BUT and cell rotation/translation (Example 2.5)

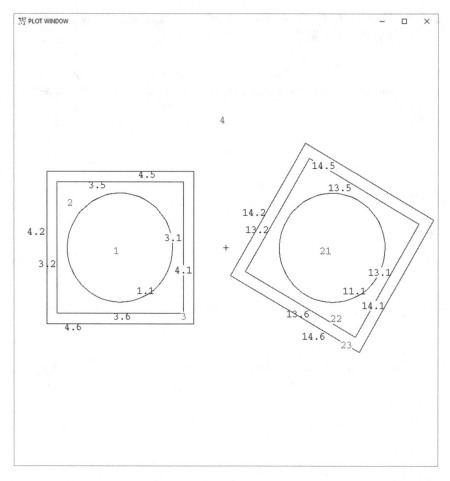

Fig. 2.5 Cylinder inside a 1 cm thick box duplicated into a second rotated box specifying 2nd box and using surface rotation/translation (Example 2.6)

Example 2.5 Cylinder Inside a 1 cm Thick Box Duplicated into a Second Rotated Box Using LIKE BUT and Cell Rotation/Translation

```
Cylinders inside 1-cm thick box
1    0    -1                   imp:n = 1  $ cylinder
2    0    1 -3                 imp:n = 1  $ inner box
3    0    3 -4                 imp:n = 1  $ outer box
4    0    4 #11 #12 #13        imp:n = 0  $ Outside world
11 like 1 but *TRCL=(20 0 0 30 90 120 90 0 90 -60 90 30) $ 2nd
cylinder
12 like 2 but TRCL=100             $ 2nd inner box
13 like 3 but TRCL=100             $ 2nd outer box
```

```
1 rcc  0 0 0 0 12 0 5   $ Right circular cylinder, 12 high, radius 5
3 rpp -6 6 0 14 -6 6    $ Right parallelepiped -6<x<6 0<y<14 -6<z<6
4 rpp -7 7 -1 15 -7 7   $ Right parallelepiped -7<x<7 -1<y<15 -7<z<7

sdef  pos=d1 erg = 1
si1 L 0 6 0    20 6 0
sp1     1         1
nps  1000000
*TR100  20 0 0   30 90 120   90 0 90    -60 90 30
```

Example 2.6 Cylinder Inside a 1 cm Thick Box Duplicated into a Second Rotated Box Specifying the 2nd Box and Using Surface Rotation/Translation

```
Cylinders inside 1-cm thick box
1    0      -1              imp:n = 1 $ cylinder
2    0  1  -3              imp:n = 1 $ inner box
3    0  3  -4              imp:n = 1 $ outer box
21   0     -11            imp:n = 1 $ 2nd cylinder
22   0 11 -13            imp:n = 1 $ 2nd inner box
23   0 13 -14            imp:n = 1 $ 2nd outer box
4    0  4  14             imp:n = 0 $ Outside world

1      rcc 0 0 0 0 12 0 5     $ Right circular cylinder, 12 high,
                                      radius 5
3      rpp -6 6 0 14 -6 6     $ Right parallelepiped -6<x<6
                                      0<y<14 -6<z<6
4      rpp -7 7 -1 15 -7 7    $ Right parallelepiped -7<x<7
                                      -1<y<15 -7<z<7
11 100  rcc 0 0 0 0 12 0 5    $ Right circular cylinder, 12 high,
                                      radius 5
13 100  rpp -6 6 0 14 -6 6    $ Right parallelepiped -6<x<6
                                      0<y<14 -6<z<6
14 100  rpp -7 7 -1 15 -7 7 $ Right parallelepiped -7<x<7
                                      -1<y<15 -7<z<7

sdef  pos=d1 erg = 1
si1 L 0 6 0 20 6 0
sp1     1       1
nps   1000000
*TR100   20 0 0    30 90 120    90 0 90    -60 90 30
```

In Example 2.5, cells 11, 12, and 13 are like cells 1, 2, and 3 but translated and rotated. The rotation translation is

$O_1\ O_2\ O_3$ XX' YX' ZX'   XY' YY' ZY' XZ' YZ' ZZ' m where X, Y, Z are the Cartesian coordinate axis vectors and X',Y',Z' are the rotated axis vectors. If $m = -1$, then the coordinate systems are reversed. Thus

```
*TR100   20 0 0    30 90 120    90 0 90    -60 90 30
```

is a 30° rotation in the XZ plane translated 20 cm in the X-direction. TR and TRCL are specified in cosine; *TR and *TRCL are specified in degrees. The transformation can be either on the cell card, TRCL=(…) as in cell 11, or on a TR card in the data section of the input and specified on the cell card by TRCL=100, as in cells 12 and 13.

```
4    0  4 #11 #12 #13    imp:n = 0  $ Outside world
```

The complement operator, #, is now required in cell 4 to indicate that LIKE BUT cells 11, 12, and 13 are excluded from the outside world in addition to everything outside surface 4.

When a cell is translated, it must create new surfaces. In Fig. 2.4, surfaces 11001, 12003, and 13004 are added for surface 1 of cell 11, surface 3 of cell 12, and surface 4 of cell 13. Like macrobody surfaces 1, 2, and 3, these surfaces have facets. Note that the Y-axis planes in cells 11, 12, and 13 are the same as those in cells 1, 2, and 3. Consequently the Y-axis facets, 1.2, 1.3, 3.3, 3.4, 4.3, and 4.4 are used in both cells 1, 2, 3 and 11, 12, 13. New facets are created only for the X-axis and Z-axis facets. The complement, #, operator is much more convenient than using the unions of the new and old named facets in the description of cell 4.

Example 2.6 duplicates the cylinder in a 1 cm thick box by simply adding cells 21, 22, 23 and surfaces 11, 13, and 14. There is no LIKE BUT or cell rotation/ translation, but now there is surface rotation/translation. Surfaces 11, 12, and 14 refer to transformation TR100 by putting the transformation number after the surface number.

2.1.7   Filled Cells: Universes

Example 2.7 illustrates universe/fill geometries. The cylinder and region outside it from Example 2.3/Example 2.4 are made into their own universe by adding $u = 1$ to cells 1, 2, and 3. Figure 2.6 shows the full geometry. It is composed of the "real world" universe, $u = 0$, geometry illustrated in Fig. 2.7. Note that there is no longer cell 4, the outside world, because universe 1 cell 3 extends out to infinity—namely everything outside surface 3, as illustrated in Fig. 2.8. Instead, cell 14 is added, which is bounded by everything inside surface 4 and filled with universe 1. Cell 11 is also filled with universe 1. Cell 15 is now the outside world.

Example 2.7 Cylinder Inside a 1 cm Thick Box Duplicated into a Second Rotated Box by Universe Fill

```
Cylinders inside 1-cm thick box
1    0    -1    u=1    imp:n=1  $ cylinder
2    0  1 -3    u=1    imp:n=1  $ inner box
3    0  3       u=1    imp:n=1  $ outside inner box
```

```
14    0     -4   fill=1    imp:n=1   $ outer box
11 like 14 but TRCL=100              $ 2nd cylinder in box
15    0   #14 #11          imp:n=0   $ outside world


1     rcc  0 0 0 0 12 0 5    $ Right circular cylinder, 12 high, radius 5
3     rpp -6 6 0 14 -6 6     $ Right parallelepiped -6<x<6 0<y<14 -6<z<6
4     rpp -7 7 -1 15 -7 7    $ Right parallelepiped -7<x<7 -1<y<15 -7<z<7

sdef  pos=0 6 0 erg = 1 cel=d2
si2 L (1<14) (1<11)
sp2       1        1
nps  1000000
PRINT
*TR100   20 0 0    30 90 120    90 0 90    -60 90 30
```

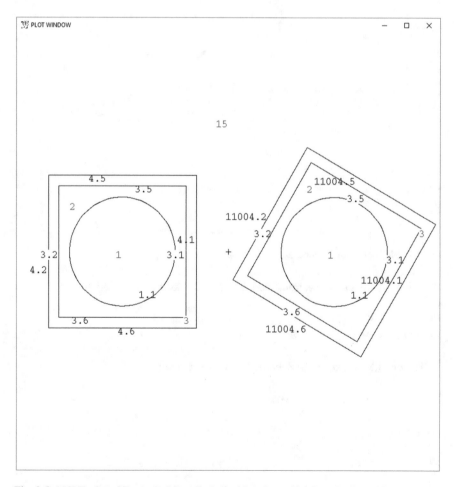

Fig. 2.6 MCNP plot of Example 2.7, cylinder inside a 1 cm thick box duplicated into a second rotated box by universe fill

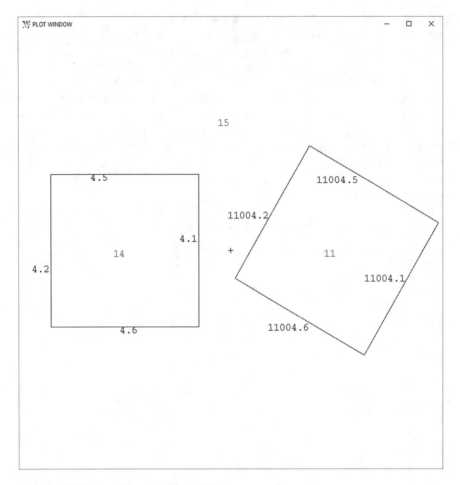

Fig. 2.7 MCNP plot of Example 2.7 level 0 (universe $u = 0$)

Cell 11 is the outer box translated 20 cm in the X-axis and tilted 30°:

```
14    0     -4   fill 1   imp:n=1   $ outer box
11 like 14 but TRCL=100            $ 2nd cylinder in box
```

The outside world can be specified either of two ways:

```
15    0   #14 #11          imp:n=0   $ outside world
```

or

```
15    0    4 #11           imp:n=0   $ outside world
```

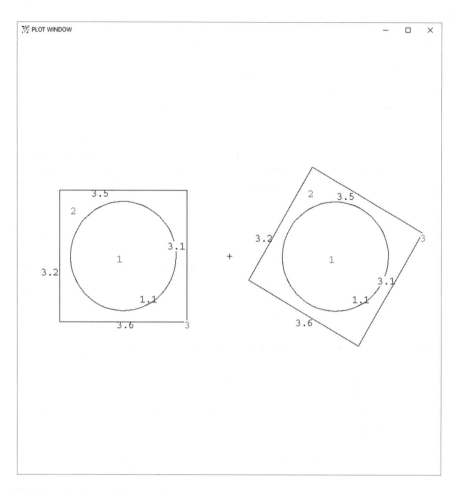

Fig. 2.8 MCNP plot of Example 2.7 level 1 (universe $u = 1$)

Cells 11, 14, and 15 in Fig. 2.6 comprise universe 0, "the real world," and $u = 0$ on the cell cards is optional. Note how the surfaces of LIKE n BUT cell 11 have been renamed 11004 for facets 1,2,5, and 6. The output file describes what happens to Y-axis top and bottom facets 3 and 4:

```
surface  1.3 and surface   3.4 are the same.    3.4 will be deleted.

surface  4.3 and surface  11004.3 are the same.  11004.3 will be deleted.

surface  4.4 and surface  11004.4 are the same.  11004.4 will be deleted.
```

The filled cells, 1, 2, and 3, retain their names, surface numbers, and coordinates. Consequently, the source must be changed to a distribution of cells:

```
sdef   pos=0 6 0 erg = 1 cel=d2
si2 L  (1<14)  (1<11)
sp2      1        1
```

The source location, pos=0 6 0, is the same for cell 1 level 1 whether it is in the universe below cell 11 or cell 14 (level 0). The first 50 histories, Print Table 100 in the output file, show the source starting in cell 1 at x, y, $z = 0$, 6, 0 all the time. The source is in either cell 11 or 14 about half the time and is at x, y, $z = 20$, 6, 0 in cell 11 and at x, y, z at 0, 6, 0 in cell 14.

```
nps    x           y          z     cell surf   u           v          w

 1  2.000E+01 6.000E+00 0.000E+00 11       8.004E-01   5.963E-01 6.178E-02
    0.000E+00 6.000E+00 0.000E+00  1  0  6.623E-01   5.963E-01 4.537E-01
 2  0.000E+00 6.000E+00 0.000E+00 14      -2.775E-01 -1.028E-01 9.552E-01
    0.000E+00 6.000E+00 0.000E+00  1  0 -2.775E-01 -1.028E-01 9.552E-01
```

Print Table 128 in the output file shows the activity in each cell and its path:

```
1neutron   activity in each repeated structure / lattice element
print table 128
           source    entering  collisions          path

           500488     500488        0        1 <    14
                0     442581        0        2 <    14
                0     500488        0        3 <    14
           499512     499512        0        1 <    11
                0     441725        0        2 <    11
                0     499512        0        3 <    11
                0          0        0       15

          1000000    2884306        0
```

Print Table 128 is sometimes turned off even when the PRINT card is in the input deck. Print Table 128 can require excessive amounts of computer time and storage, locating the path to thousands of elements at lower universe levels such as in the case of a nuclear reactor model or human phantom model consisting of millions of pixels. When the MCNP code turns off Print Table 128, the following warning is issued:

```
warning.  universe map (print table 128) disabled.
```

2.1.8 Lattice Geometries

Example 2.8 illustrates square and hexagonal lattices and is shown in Fig. 2.9. The geometry consists of an RPP box (cell 14) filled with universe 11 and another one like it but translated 20 cm in the *X*-direction and filled with universe 12, as shown in Fig. 2.10. Both are contained in a huge sphere, surface 20.

Example 2.8 Square and Hex Lattice File

```
Square and hex lattices
1   313 -10.44      -1          u=1    imp:n=1  $ UO2
2   201  -6.55      -2 1        u=1    imp:n=1  $ Zr clad
3   609  -1.00       2          u=1    imp:n=1  $ H2O
21   0        -31 lat=1 fill=1  u=11   imp:n=1  $ Square lattice
22   0        -32 lat=2 fill=1  u=12   imp:n=1  $ Hexagonal lattice
14   0         -4             fill 11  imp:n=1  $ Square lattice box
11 like 14 but TRCL=(20 0 0) fill 12            $ Hexagonal lattice box
15   0   #14 #11 -20                   imp:n=1  $ outside boxes
16   0               20                imp:n=0  $ outside world

1    rcc  0 0 0  0 12 0  .5        $ Right circular cylinder
2    rcc  0 0 0  0 12 0  .6        $ Right circular cylinder
4    rpp -7 7    0 12  -7 7        $ Right parallelepiped 14x12x14
31   rpp -1 1    0 12  -1 1        $ Right parallelepiped 2x12x2
32   rhp  0 0 0  0 12 0  1 0 0     $ Right hex 12x1
20   sph 0 0 0 50                  $ Sphere containing lattice boxes

m201    40090 -0.505239    40091 -0.110180    40092 -0.168413
        40094 -0.170672    40096 -0.027496    nlib=80c
m313   92235.80c -0.02759  92238.80c -0.85391  8016.80c -0.11850
M609    1001.80c 2   8016.80c 1
MT609   LWTR
sdef   pos=10 6 0
nps   1000000
PRINT
```

At universe level 1, universe 11 fills cell 14 and universe 12 fills cell, as illustrated in Fig. 2.11. Cell 21 of universe 11 is a square lattice, LAT = 1, and cell 22 of universe 12 is a hexagonal lattice, LAT = 2.

Figure 2.12 illustrates universe level 2, where both the rectangular and hexagonal lattices are filled with universe 1. Universe 1 is a UO_2 fuel rod (cell 1, material M313) in Zr cladding (cell 2, material M201) in an infinite extent of water (cell 3, material 609) outside surface 2.

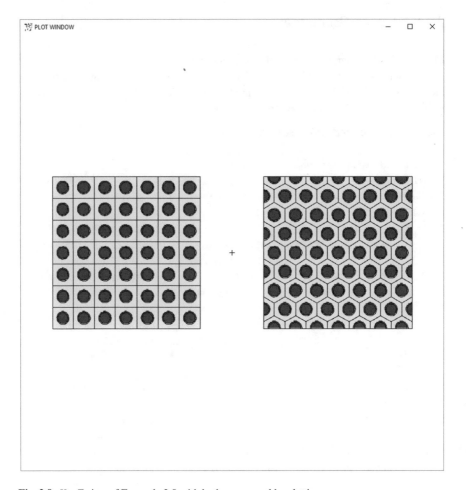

Fig. 2.9 $X - Z$ view of Example 2.8 with both square and hex lattices

The fuel rods can be differentiated according to which lattice element they fill. The indices are plotted in Fig. 2.13 using the LA=0 1 IJK command. With the interactive plotter, click IJK on the right margin and then label L2 in the lower left menu box.

The source of Example 2.8 is a 14 MeV isotropic neutron (default) midway between the two lattice boxes at POS = 10 6 0. From the problem summary in the output file, 0.13 fission neutrons and 0.07 (n,xn) neutrons are produced per source particle.

Print Table 128 in the output file shows the activity in each cell and its path:

```
1neutron   activity in each repeated structure / lattice element
print table 128
```

source	entering	collisions		path		
0	708051	216431	1 <	21 <	14	
0	1587248	67168	2 <	21 <	14	
0	2736613	1330851	3 <	21 <	14	
0	921642	251038	1 <	22 <	11	
0	1868093	79206	2 <	22 <	11	
0	2972581	1227971	3 <	22 <	11	
1000000	1801413	0	15			
0	0	0	16			
1000000	12595641	3172665				

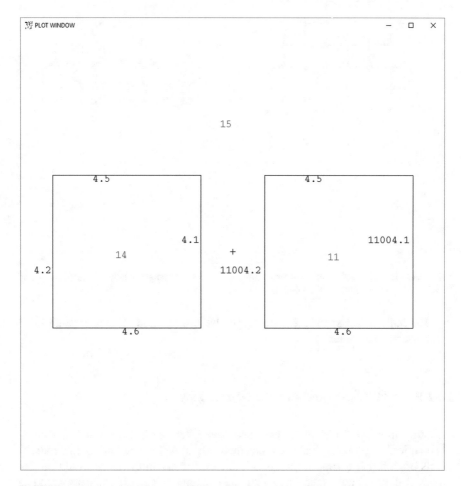

Fig. 2.10 Example 2.8 level $u = 0$ universe. Note how surface facets 11004.1 and 11004.2 are created because cell 11 is like cell 14 but surfaces 4.1 and 4.2 remain with cell 14

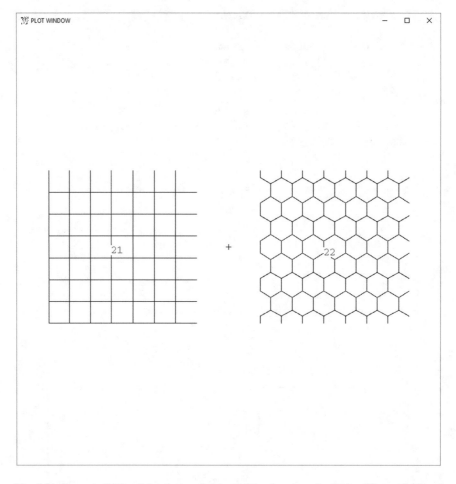

Fig. 2.11 Example 2.8 level 1 universe showing LAT = 1 rectangular lattice filling cell 21 and LAT = 2 hexagonal lattice filling cell 22

The lattices in Example 2.8 are simple lattices because all elements are filled with the same universe.

2.1.9 Fully Specified Lattice Geometries

A lattice in which the elements are filled with different universes is a "fully specified lattice." In a reactor fuel assembly, these would be for water holes, fuel rods, or fuels of different compositions. In a ^{3}He detector assembly, these would be ^{3}He tubes of different atmospheres. In a human phantom model, the elements would be

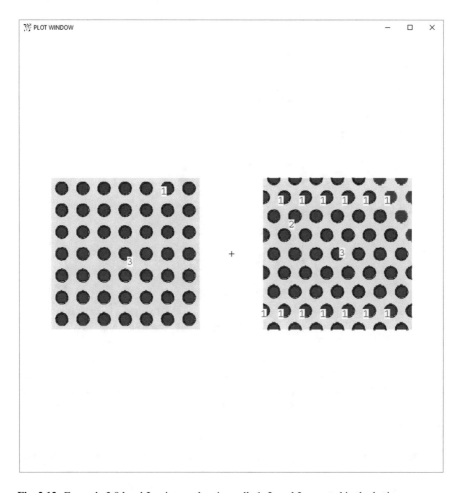

Fig. 2.12 Example 2.8 level 2 universe showing cells 1, 2, and 3 repeated in the lattices

filled with different human organs. Example 2.9 is an MCNP input file with a fully specified lattice geometry.

Example 2.9 Fully Specified Square and Hex Lattices with Multiple Universes

```
Square and hex lattices
1   313 -10.44     -1        u=1    imp:n=1  $ UO2
2   201  -6.55     -2 1      u=1    imp:n=1  $ Zr clad
3   609  -1.00      2        u=1    imp:n=1  $ H2O
21 111  -4.00     -31 lat=1 u=11  imp:n=1  $ Square lattice
              fill = -3:3 0:0 -3:3
                    1 1 1 1 1 1 1
```

```
                         1 1 1 1 1 1 1
                         1 1 1 1 1 1 1
                         1 1 1 1 1 1 1
                         1 1 1 1 1 1 1
                         1 1 1 1 1 1 1
                        11 1 1 1 1 1 1
22 111   -4.00    -32 lat=2 u=12   imp:n=1  $ Hexagonal
             fill = -5:5 -4:4 0:0
                       1 1 1 1 1 1 1 1 1 1 1
                       1 1 1 1 1 1 1 1 1 1 1
                       1 1 1 1 1 1 1 1 1 1 1  .
                       1 1 1 1 1 1 1 1 1 1 1
                       1 1 1 1 1 1 1 1 1 1 1
                       1 1 1 1 1 1 1 1 1 1 1
                       1 1 1 1 1 1 1 1 1 1 1
                       1 1 1 1 1 1 1 1 1 1 1
                      12 1 1 1 1 1 1 1 1 1 1
14   0     -4              fill 11 imp:n=1  $ Square lattice box
11 like 14 but TRCL=(20 0 0) fill 12       $ Hexagonal lattice box
15   0   #14 #11 -20            imp:n=1  $ outside boxes
16   0              20          imp:n=0  $ outside world

1   rcc  0 0 0  0 12 0  .5     $ Right circular cylinder, 12h,.5r
2   rcc  0 0 0  0 12 0  .6     $ Right circular cylinder, 12h,.6r
4   rpp -7 7   0 12  -7 7      $ Right parallelepiped 14x12x14
31  rpp -1 1   0 12  -1 1      $ Right parallelepiped 2x12x2
32  rhp  0 0 0  0 12 0  1 0 0  $ Right hex 12x1
20  sph 0 0 0 50              $ Sphere containing lattice boxes

m201   40090 -0.505239   40091 -0.110180    40092 -0.168413
       40094 -0.170672   40096 -0.027496   nlib=80c
m313   92235.80c -0.02759   92238.80c -0.85391   8016.80c -0.11850
M609   1001.80c 2  8016.80c 1
MT609  LWTR
M111   6000 1  1001 1  7014 1
sdef   pos=10 6 0
nps  1000000
PRINT
```

The simple lattice definition of Example 2.8 was

```
21   0   -31 lat=1 fill=1 u=11  imp:n=1  $ Square lattice
22   0   -32 lat=2 fill=1 u=12  imp:n=1  $ Hexagonal lattice
```

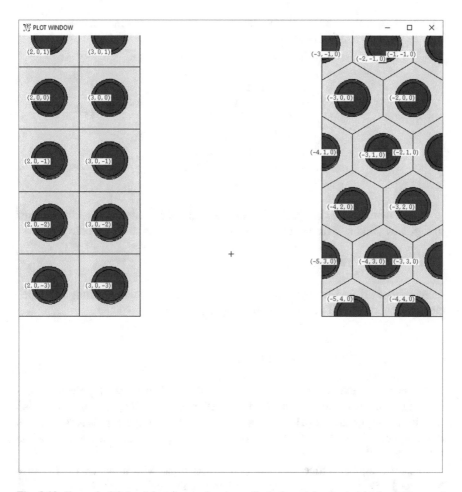

Fig. 2.13 Example 2.8 level 2 universe showing cells 1, 2, and 3 repeated in the lattices and labeled by lattice index

To change one element the fully specified lattice specification is

```
21 111  -4.00   -31 lat=1 u=11  imp:n=1  $ Square lattice
            fill = -3:3 0:0 -3:3
                   1 1 1 1 1 1 1
                   1 1 1 1 1 1 1
                   1 1 1 1 1 1 1
                   1 1 1 1 1 1 1
                   1 1 1 1 1 1 1
                   1 1 1 1 1 1 1
                  11 1 1 1 1 1 1
```

```
22 111   -4.00    -32 lat=2 u=12  imp:n=1  $ Hexagonal lattice
              fill = -5:5 -4:4 0:0
                         1 1 1 1 1 1 1 1 1 1 1
                         1 1 1 1 1 1 1 1 1 1 1
                         1 1 1 1 1 1 1 1 1 1 1
                         1 1 1 1 1 1 1 1 1 1 1
                         1 1 1 1 1 1 1 1 1 1 1
                         1 1 1 1 1 1 1 1 1 1 1
                         1 1 1 1 1 1 1 1 1 1 1
                         1 1 1 1 1 1 1 1 1 1 1
                        12 1 1 1 1 1 1 1 1 1 1
```

A new material, M111, is added to cells 21 and 22 so that the different elements in Fig. 2.14 contain different universes and are plotted in a different color.

```
M111   6000 1   1001 1   7014 1
```

Instead of FILL=1 for both lattices, the FILL is replaced with the range of lattice elements needed to fill the filling cells 21 and 22:

```
              fill = -3:3 0:0 -3:3
              fill = -5:5 -4:4 0:0
```

In the square lattice, element [−3,0,3] is filled with universe 11. Because universe 11 is the same universe as its lattice cell, cell 21 $u = 11$, the lattice is "filled with itself" with material M111 instead of the fuel pins of universe 1. Note that this is the upper left in the MCNP plot but the lower left in the lattice map of the FILL entry.

In the hexagonal lattice, element [−5,4,0] is filled with universe 12. Because universe 12 is the same universe as its lattice cell, cell 22 $u = 12$, the lattice is "filled with itself" with material M111 instead of the fuel pins of universe 1. Note that unlike the rectangular lattice, the universe 12 element is the lower left in both the MCNP plot and in the lattice map of the FILL entry. In the hexagonal lattice there are two constraints: the first and second indices must increase across adjacent surfaces and the third index must increase in one or the other direction along the length of the hexagonal axis. A rectangular lattice can completely fit into a rectangular box, but a hexagonal lattice cannot completely fit in a rectangular box or any other MCNP microbody surface shape. Consequently, the hexagonal lattice of Fig. 2.14 is filled with 99 elements −5:5 −4:4 0:0, but only 45 are full hexes, 22 are cut in half, and the remaining 32 are cut off altogether. A further description of the MCNP LAT and FILL capability is in the MCNP Manual [2] Section 3.3.1.5.2 and 3.3.1.5.3.

The lattice elements are most easily recognizable from geometry plots that have the label LA=0 1 IJK or by clicking on ijk in the right margin and then L2 in the lower left menu. These indices are shown in Figs. 2.15 and 2.16.

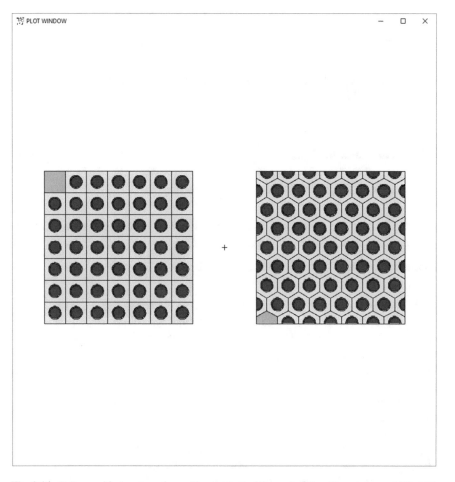

Fig. 2.14 Fully specified rectangular and hex lattices of Example 2.9, with one element filled differently in each

The fully specified hexagonal lattice is more commonly specified in the MCNP input file with the elements of the hexagonal lattice cut off by the filling cell (cell 11 filled with universe 12, which is the hex lattice cell) being filled with universe 0:

```
22 111   -4.00    -32 lat=2 u=12   imp:n=1  $ Hexagonal
         fill = -5:5 -4:4 0:0
                         0 0 0 0 1 1 1 1 1 1 1
                          0 0 0 1 1 1 1 1 1 1 1
                           0 0 0 1 1 1 1 1 1 1 0
                            0 0 1 1 1 1 1 1 1 1 0
                             0 0 1 1 1 1 1 1 1 0 0
                              0 1 1 1 1 1 1 1 1 0 0
```

```
             0 1 1 1 1 1 1 1 0 0 0
             1 1 1 1 1 1 1 1 0 0 0
             12 1 1 1 1 1 1 0 0 0 0
```

The above hexagonal lattice is also used later in Example 2.11, Example 2.12, and Example 2.19.

2.2   Materials and Cross Sections

2.2.1   *Specifying Materials*

An MCNP input file consists of three sections: cells, sources, and data. The data section of Example 2.8 is shown in Example 2.10:

Example 2.10 Data Cards from Example 2.6

```
m201    40090 -0.505239    40091 -0.110180    40092 -0.168413
        40094 -0.170672    40096 -0.027496    nlib=80c
m313    92235.80c -0.02759    92238.80c -0.85391    8016.80c -0.11850
M609    1001.80c 2   8016.80c 1
MT609   LWTR
sdef    pos=10 6 0
nps     1000000
PRINT
```

Materials are specified on the materials cards, but additional input in the data section is often required. Material cards consist of ZAID-fraction pairs. Negative fractions are mass fractions; positive fractions are atom fractions. The fractions are normalized to unity for all materials. The ZAID numbers consist of XXXAAA.nnx, where ZZZ is the atomic number, AAA is the atomic mass, nn is the specific evaluation, and x is the data type.

Material numbers are arbitrary. For material m201, ZAID = 40090 indicates $Z = 40$ zirconium isotope $A = 90$. Thus, material 201 consists of five zirconium isotopes—^{90}Zr, ^{91}Zr, ^{92}Zr, ^{94}Zr, and ^{96}Zr—with mass fractions 0.505239, 0.110180, 0.168413, 0.170672, and 0.027496. The optional nlib=80c sets nnx for each nuclide to 80c, which, in this case, indicates ENDF/B-VII.1 continuous-energy neutron cross sections.

Material 313 is specified as uranium dioxide, UO_2, 3.17 at% or 3.13 wt% enriched in ^{235}U.

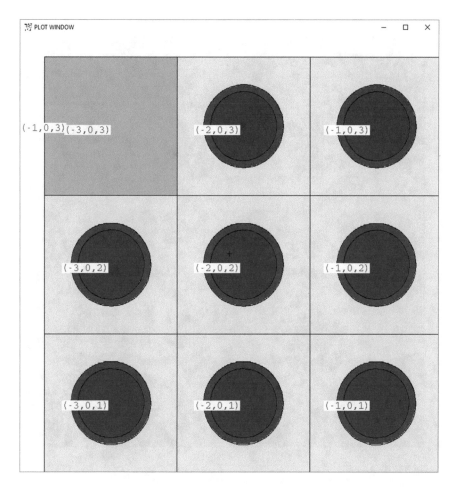

Fig. 2.15 Square lattice with ijk lattice indices. The extra (−1,0,3) in the upper left is an MCNP6.2 plotting bug that repeats the label of the upper right. The I indices increase left to right, which is the X-direction for RPP surface 31 of lattice cell 21. The J indices are from 0:0 and they increase in the Y-direction (out of the plane of the plot) for RPP surface 21 of lattice cell 21. The K indices increase from bottom to top, which is the X-direction for RPP surface 31 of lattice cell 21

2.2.2   Neutron Cross Sections

Cross sections are available only for common isotopes. If the cross section for an isotope specified on the material card does not exist, then the MCNP code will terminate with a fatal error. For example, if there is no tritium neutron transport cross section available,

```
M609    1003.80c 2  8016.80c 1
```

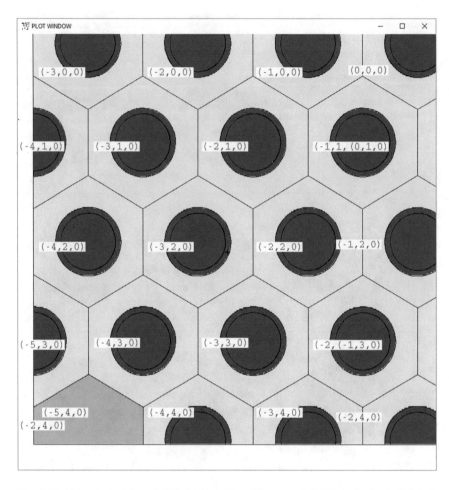

Fig. 2.16 Hexagonal lattice with ijk lattice indices. The extra (−2,4,0) in the lower left is an MCNP6.2 plotting bug that repeats the label of the lower right. The I indices increase at 60° from lower left to upper right. The *J* indices increase at 120° from upper left to lower right. The *K* indices increase from bottom to top of the hex axis—from 0:0 in this case—because the hex is only one element high

then the code terminates with the message

```
fatal error.  cross-section tables missing for zaid =  1003.80c
```

Even worse, in problems with protons or other charged particles or ions, missing cross-section data will cause the MCNP code to use theoretical models for those data instead. The code will also use model physics for energies higher than the highest tabulated energy in any cross-section library it is using. These models are good for neutrons with energies >150 MeV and particles other than neutrons; the models are generally poor for neutrons with energies <150 MeV. It is important to check the cross sections used to ensure that they are in the range of particle energy simulated in each problem.

Further, different cross-section evaluations are available for different computers and different MCNP versions.

Wrong cross sections give wrong answers—the most frequent user error. There is a document listing the most available data [4], but these are not necessarily the ones available to the MCNP code on a particular computer and code installation. Consequently, the MCNP code frequently selects unexpected cross sections; therefore, it is *imperative* to always *check* in the MCNP OUTP output files to see which cross sections are actually being used by the code. The cross sections used are described in Print Table 100:

```
40090.80c 96741 Zr90 ENDF71x (jlconlin) Ref. see jlconlin (ref 09/10/2012)
mat4025  12/18/12
        Energy range:   1.00000E-11  to  2.00000E+01 MeV.
        particle-production data for protons  being expunged from  40090.80c
        particle-production data for deuterons being expunged from  40090.80c
        particle-production data for tritons  being expunged from  40090.80c
        particle-production data for helions  being expunged from  40090.80c
        particle-production data for alphas   being expunged from  40090.80c
        probability tables used from 5.3500E-02 to 1.0000E-01 mev.
```

The cross section used in this case is ^{90}Zr from an experimental ENDF/B-VII library, ENDF71x, and in the energy range up to 20 MeV. Whereas the problem runs only neutrons, the production data for protons, deuterons, tritons, helions (^{3}He), and alphas are deleted to save storage; these reactions are treated as capture. The probability tables from 53.5 to 100 keV provide better physics in the unresolved cross-section energy range. Often the temperature of the cross-section data is also provided.

Figure 2.17 is an MCNP plot of material 313. The MCNP code is run in the cross-section plotting mode:

```
MCNP6  I=Ex11_6a  name=junk.  IXZ
```

and the MCNP plot command is

```
xs m313 cop xs 92238.80c cop xs 92235.80c
```

MCNP cross-section plotting is further described in Sect. 2.5. The cross-section table starts at 1e−11 MeV (1e−5 eV); the data are horizontally extrapolated at lower energies, though all neutrons are essentially absorbed before their energies become that low. From 1e−11 to 1e−5 MeV the neutron cross-section has a 1/v tail, where v is the neutron velocity. Above that is the resonance cross-section region up to

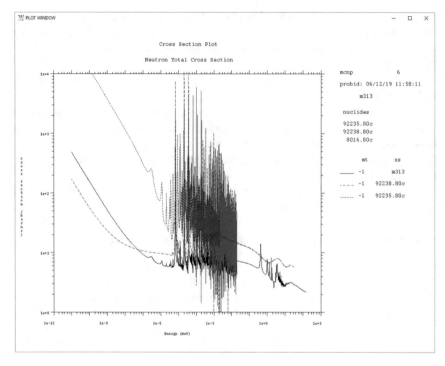

Fig. 2.17 Neutron cross section of material M313, UO$_2$ (black), ^{238}U (blue), and ^{235}U (red) in Example 2.10

2 keV for ^{235}U and 20 keV for ^{238}U. Above that is the unresolved resonance cross-section region—where the resonances are so close together that they cannot be resolved and are represented by either a smooth curve or probability tables. Above 500 keV are inelastic and reaction cross sections such as secondary particle production. The uranium cross sections end at 20 MeV, and physics models are used above, or the cross section is merely extrapolated. The material m313 cross section goes up to 150 MeV because of the oxygen cross section.

2.2.3   Low-Energy Neutron Problems: Thermal Free Gas Treatment

Neutrons at very low energies and all collisions with hydrogen and very low-Z nuclides are affected by the thermal motion of the nuclei with which they interact. MCNP treats the kinematics of neutron scatter with moving nuclei with the thermal free gas treatment.

The thermal free gas treatment assumes room temperature. For problems at other temperatures, the temperature, kT, in MeV units must be specified.

kT (MeV)	Unit of T
$8.617 \times 10^{-11}\, T$	Kelvin
$8.617 \times 10^{-11}\, (T + 273.15)$	degrees C
$4.787 \times 10^{-11}\, T$	degrees R
$4.787 \times 10^{-11}\, (T + 459.67)$	degrees F

The temperature is usually specified on cell cards (using TMPn cell parameter). For cells 1 and 2 in Example 2.10, assume the temperatures are 300 °C and 600 °C. Then

```
1   313 -10.44     -1      u=1    imp:n=1  $ UO2
2   201  -6.55     -2 1    u=1    imp:n=1  $ Zr clad
```

becomes

```
1   313 -10.44     -1      u=1   tmp1=4.939e-8 imp:n=1  $ UO2
2   201  -6.55     -2 1    u=1   tmp1=7.524e-8 imp:n=1  $ Zr clad
```

Room temperature (20.5 °C = 2.5301e−8 MeV) is assumed for all cells without a TMP1 entry. Temperatures may be time-dependent using the THTME card and TMP1, TMP2, TMP3, etc.

Alternatively, temperatures may be specified in the data section of input for the first two cells among the cell cards and using the default (room temperature) for the other 7 cells in the problem.

```
TMP1   3.216e-8   4.939e-8   7J
```

Temperatures may also be time-dependent:

```
THTME   1e8 6e9
TMP1   3.216e-8   4.939e-8   7J
TMP2   4.939e-8   7R
```

The above example has 100 °C and 300 °C for the first two cells and room temperature for the remaining seven cells from time 0 to 1 s (1e8 shakes). The temperature is 300 °C for all cells at 60 s (6e9 shakes) and later. The temperature is linearly interpolated from 1 s < time < 60 s. Note the use of J (jump) notation to take the default values and the R (repeat) notation to repeat the temperatures at 60 s.

Cell temperatures impact elastic scattering cross sections and collisions kinematics. The temperature on the TMP card does not affect inelastic or reactions cross sections, thermal scattering kernels, or resonances in scattering. Cross-section resonance broadening is done only by choosing cross-section data files at the desired temperature.

2.2.4   Low-Energy Neutron Problem Data: S(α,β)
Thermal Treatment

Molecular binding of nuclides into molecules also affects the scattering of neutrons
at low energies (few eV). The MCNP code treats these scatters with the S(α,β) thermal treatment [2]. By default, the S(α,β) treatment is off because the code does not
know the compositions of molecules. The MCNP input must explicitly turn it on. If
low-energy neutrons are important (almost always the case if fission is present),
failure to specify S(α,β) will result in wrong answers. There is almost no computer
time or storage cost for including S(α,β), so it is recommended to specify S(α,β) for
all neutron problems where the limited amount of S(α,β) data is available.

```
M609   1001 2   8016 1
MT609 LWTR
```

As illustrated in Example 2.10, material M609 is light water, so its S(α,β) molecule is specified on the above MT609 card.

S(α,β) data are available for different temperatures of a limited number of molecules of hydrogen, deuterium, beryllium, graphite, oxygen, polyethylene, benzene,
and more. The most reliable way to tell what is available is to examine the XSDIR
data-directory file that comes with each version of the MCNP code and then to
verify what the MCNP code is actually used by examining the output file. Further
documentation is also available [2, 4].

As another example, consider material 111 from Example 2.7:

```
M111   6000.80c 1   1001.80c 1   7014.80c 1
MT111 POLY.10T BENZ.10T GRPH.10T
```

By adding the MT111 card to M111 in Example 2.7, the nitrogen, 7014.80c, will
have no S(α,β) because none of the MT111 molecules contains nitrogen data. The
hydrogen, 1001.80c, will use POLY.10T for S(α,β) because POLY appears before
BENZ on the MT111 card and is the first MT111 molecule containing hydrogen
data. Carbon, 6000.80c, will use GRPH.10T because POLY and BENZ apply only
to the hydrogen in POLY and BENZ, not the carbon; GRPH is the only MT111
molecule containing carbon data. The carbon, 6000.80c, and graphite S(α,β),
GRPH.10T, are plotted in Example 2.11.

Figure 2.18 is an MCNP plot of material M111 in Example 2.7. MCNP is run in
the cross-section plotting mode:

```
MCNP6  I=Ex12_1b  name=junk.  IXZ
```

The MCNP plot command is

```
xlim 1e-11 1e-5 xs 6000.80c mt 1 cop xs grph.10t mt 1 cop mt 2
cop mt 4
```

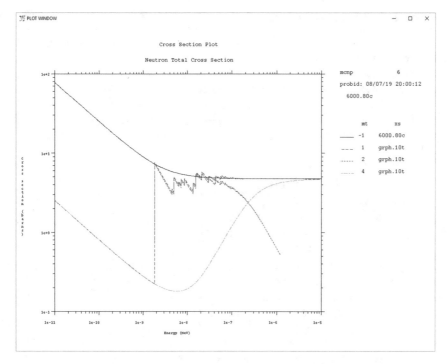

Fig. 2.18 Carbon 6000.80c continuous-energy total (MT = −1) neutron cross section (black) co-plotted with GRPH.10T S(α,β) cross-section total (MT = 1) (blue), elastic (MT = 2) (red), and inelastic (MT = 4) (green)

MCNP cross-section plotting is further described in Sect. 2.5. The S(α,β) cross sections extend from about 1e−11 MeV to the S(α,β) cutoff, generally about 1e−5 MeV. Above the cutoff, only the continuous-energy data are used. Below the cutoff, only the S(α,β) data are used. S(α,β) data should be used whenever possible. When neutrons do not scatter below the cutoff, the S(α,β) data are ignored. When neutrons do scatter below the cutoff, then the molecular thermal effects are important.

2.2.5   Photon Cross Sections

The MODE card is required whenever anything other than a neutron is transported. In Example 2.11, the M0 causes all neutrons to use the .70c ENDF/B-VII data and all photons to use the .04p photon data. If NLIB or PLIB were missing, then MCNP would use the default data libraries, which is whatever libraries MCNP finds first on the particular computer for the particular version of MCNP. Generally the default photon cross sections are good enough, but the output file should always be checked to see if the ones selected are appropriate.

Example 2.11 Photon Problem

```
Square and hex lattices
1   313 -10.44      -1          u=1    imp:n=1   $ UO2
2   201  -6.55      -2 1        u=1    imp:n=1   $ Zr clad
3   609  -1.00       2          u=1    imp:n=1   $ H2O
21  111  -4.00      -31 lat=1 u=11    imp:n=1   $ Square lattice
                fill = -3:3 0:0 -3:3
                            1 1 1 1 1 1 1
                            1 1 1 1 1 1 1
                            1 1 1 1 1 1 1
                            1 1 1 1 1 1 1
                            1 1 1 1 1 1 1
                            1 1 1 1 1 1 1
                           11 1 1 1 1 1 1
22  111  -4.00      -32 lat=2 u=12    imp:n=1   $ Hexagonal
            fill = -5:5 -4:4 0:0
                        0 0 0 0 1 1 1 1 1 1 1
                         0 0 0 1 1 1 1 1 1 1 1
                          0 0 0 1 1 1 1 1 1 1 0
                           0 0 1 1 1 1 1 1 1 1 0
                            0 0 1 1 1 1 1 1 1 0 0
                             0 1 1 1 1 1 1 1 1 0 0
                              0 1 1 1 1 1 1 1 0 0 0
                               1 1 1 1 1 1 1 1 0 0 0
                               12 1 1 1 1 1 1 0 0 0 0
14   0      -4               fill  11 imp:n=1  $ Square lattice box
11 like 14 but TRCL=(20 0 0) fill 12        $ Hexagonal lattice box
15   0    #14 #11 -20              imp:n=1  $ outside boxes
16   0               20           imp:n=0  $ outside world

1   rcc  0 0 0  0 12 0  .5      $ Right circular cylinder,12h,.5r
2   rcc  0 0 0  0 12 0  .6      $ Right circular cylinder,12h,.6r
4   rpp -7 7    0 12  -7 7      $ Right parallelepiped 14x12x14
31  rpp -1 1    0 12  -1 1      $ Right parallelepiped 2x12x2
32  rhp  0 0 0  0 12 0  1 0 0   $ Right hex 12x1
20  sph 0 0 0 50               $ Sphere containing lattice boxes

MODE N P
m0 nlib=.70c plib=.04p
m201    40090 -0.505239    40091 -0.110180    40092 -0.168413
        40094 -0.170672    40096 -0.027496    nlib=80c
m313    92235 -0.02759   92238 -0.85391    8016 -0.11850
M609    1001 2   8016 1
MT609   LWTR
```

```
M111   6000 1   1001 1   7014 1
MT111 POLY BENZ GRPH
sdef   pos=10 6 0
nps   1000000
PRINT
```

Most photon reactions are photoatomic and involve interactions with the atom's electrons. Consequently, photoatomic data are elemental rather than isotopic. The MCNP code will convert

```
m201   40090 -0.505239   40091 -0.110180   40092 -0.168413
       40094 -0.170672   40096 -0.027496
m313   92235 -0.02759    92238 -0.85391   8016 -0.11850
M609   1001 2   8016 1
M111   6000 1   1001 1   7014 1
```

to

```
m201   40000 -0.505239   40000 -0.110180   40000 -0.168413
       40000 -0.170672   40000 -0.027496
m313   92000 -0.02759    92000 -0.85391   8000 -0.11850
M609   1000 2   8000 1
M111   6000 1   1000 1   7000 1
```

Note that the Zr in material M201 will still be sampled for five different mass fractions of the same Zr element 40,000; uranium in material M313 will be sampled for two different mass fractions for 92,000.

Figure 2.19 is an MCNP plot of hydrogen, oxygen, zirconium, and uranium total photon cross sections. MCNP is run in the cross-section plotting mode:

```
MCNP6   I=Ex12_2a   name=junk.   IXZ
```

and the MCNP plot command is (see section 5.3.3 of the MCNP manual):

```
ylim .01 1e7 par p mt -5 xs 1000.04p cop xs 8000.04p &
cop xs 40000.04p cop xs 92000.04p
```

Photons are considered absorbed and no longer transported below 1 keV. Photon cross sections go up to 100 MeV. The five photoatomic reactions in MCNP are (1) incoherent (Klein–Nishina Compton scatter plus form factors); (2) Coherent (Thomson scatter plus form factors); (3) photoelectric (with k- and l-shell fluorescence); (4) pair production, in which the positron immediately decays into two photons; and (5) bremsstrahlung, generated by electrons. The first four reactions are treated rigorously, but bremsstrahlung is approximate unless electrons are also transported. Electron transport is a thousand times slower than photon transport and

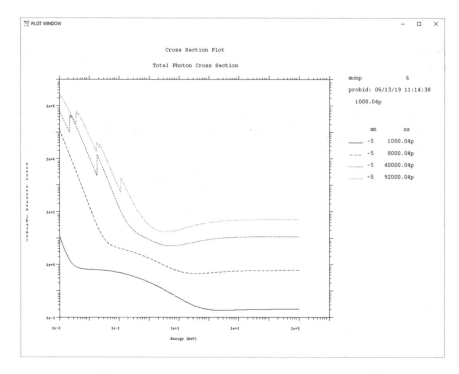

Fig. 2.19 Total photon cross section of H, O, Zr, and U

is approximate. Thus, the default photon treatment is the thick-target bremsstrahlung (TTB) approximation. TTB produces all of the photons that would be produced by full electron transport at the point of interaction and in the direction of the photoatomic collision without transporting the electrons. In a thick geometric region, this is a good approximation: after many electron interactions, the electrons remain in the same region, move insignificantly, and have essentially random directions. In a thin geometric region where the electrons lose only a few percent of their energy while slowing down, TTB is a poor approximation and a full electron transport is required.

Figure 2.20 is an MCNP plot of the basic uranium photon cross sections. MCNP is run in the cross-section plotting mode:

```
MCNP6   I=Ex12_2a   name=junk.   IXZ
```

and the MCNP plot command is

```
ylim .01 1e7 par p xs 92000.04p mt -5 cop mt -1 cop mt -2 cop
mt -3 cop mt -4
```

Bremsstrahlung is not included in the photon libraries because it is generated by electron models.

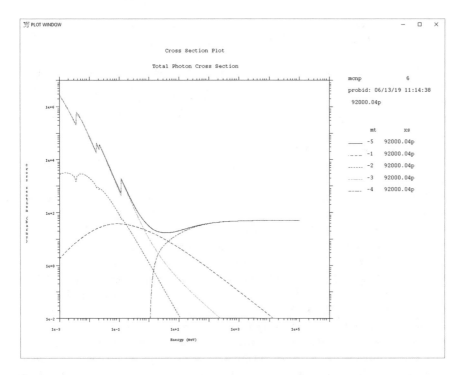

Fig. 2.20 Photon cross sections for uranium: total (MT = −5), incoherent (MT = −1), coherent (MT = −2), photoelectric (MT = −3), and pair production (MT = −4)

Photonuclear reactions are those in which photons interact with the nucleus. Because they are generally low probability, the MCNP code ignores them unless specifically requested with a PHYS:P card. Photonuclear reactions produce neutrons, photons, and other particles, and a limited number of isotopic photonuclear cross sections are available with MCNP. The ZAID number ends in a u, as in 6012.70u. Photonuclear reactions should be modeled only if secondary photonuclear particles are of particular interest.

2.2.6   Electron-Stopping Powers for Coupled Photon and Electron Problems

Example 2.11 becomes a coupled photon-electron problem by replacing the MODE card and M0 data cards:

```
MODE   P   E
M0   PLIB=04p ELIB=03e ESTEP=10 COND GAS
```

In the above example, photons use the .04p data, electrons use the .03e data, and the material is treated electrically as a gaseous conducting material. ESTEP will be described soon.

Coupled electron-photon transport is possible below 10 keV with the single-event electron transport method by specifying data from the EPRDATA14 photon library (.14p) and setting the 15th entry on the PHYS:E card. The single-event electron-photon transport models every interaction of the electron and photon random walk and is very slow and approximate. Generally, the condensed history method is used, and photons and electrons below 1 keV are simply absorbed.

Electron transport in MCNP for energies $E > 1$ keV utilizes the condensed history method.

The condensed history method models all interactions of an electron over an energy substep. An electron passing through matter will interact with each atom along its trajectory via ionization, angular scattering, and production of secondaries. Modeling this interaction exactly by the single-event model is time prohibitive. The condensed history algorithm attempts to average the effect of all of these interactions into aggregate quantities, which can be approximated over a series of major steps and several substeps. The major steps are controlled by EFAC, the 14th entry on the PHYS:E card. Each step is at an energy EFAC = 0.917 times the energy of the previous (higher) step, which provides eight energy steps for a factor of two energy loss. Setting EFAC = 0.99 makes each step at an energy 0.99 of the previous step, which provides a much finer and more accurate energy grid but increases computational time by a factor of about ten. The substeps are controlled by ESTEP on the materials card and other factors such as the particular material and energy. The default is ESTEP = 3, or 3 substeps per major or full step. The step size can be different for each material.

Here, ESTEP = 10 specifies 10 substeps for each full step. M0 applies this to all materials for which ESTEP is not otherwise specified. The other parameters specify that photons use the .04p data, electrons use the .03e data, and the material is treated electrically as a gaseous conducting material.

The highest energy, EMAX, is the 1st entry on the PHYS:E card (100 MeV by default). An electron starting at some energy E will use the electron data (stopping powers, slowing-down parameters, etc.) at its energy range and travel the distance needed to lose a substep's worth of energy. When it loses a full step of energy, or 1.0−0.917 = 8.3% of its energy, it will then use the data of that step. For example, by default, an electron at 1 MeV will use the 1 MeV data until its energy falls below 917 keV. It will then use these data until its energy falls below 814 keV, etc. In each of these ranges, it will have three substeps that determine how far the electron travels and its deflection direction. At each step, bremsstrahlung photons, photon-induced electrons, electron induced X-rays, and knock-on electrons are produced and these production mechanisms may be biased with the PHYS:E card.

Electron transport is approximately 1000 times slower than photon transport. Changing ESTEP from 3 to 20 slows down electron transport by another factor of 10. Changing EFAC to the maximum resolution of EFAC = 0.99 on the PHYS:E card slows down electron transport by another factor of 10. But the default values of

ESTEP = 3 and EFAC = 0.917 may result in too coarse an energy grid structure and wrong answers. If electron transport is unimportant, then the thick-target brems-strahlung approximation should be sufficient. If the calculation is sensitive to the electron transport, then we recommend checking if higher resolution (increasing EFAC and ESTEP) affects results. Turning off knock-on electrons (RNOK=0, 8th entry) will often speed calculations by a factor of seven without adverse effects. The effect of knock-on electrons is problem dependent. It is recommended to run first with knock-on electrons turned on (default) to include knock-on effects; then turn off knock-ons to see if tallies are unaffected and the calculation runs faster.

Positron and electron physics are identical except when they stop (fall below energy cutoff). Electrons deposit energy, whereas positrons generate annihilation photons. At high energies, positrons behave like electrons. At low energies (<1 MeV), stopping powers, bremsstrahlung, knock-ons, and annihilations are increasingly poor. Positron sources are allowed:

```
SDEF par = -e
SDEF par = f
```

Positrons may be tallied separately:

```
FTn ELC 3
```

The electron-stopping powers and other parameters are printed for each energy step and each material in the problem. We recommend printing this information in the output file only when changing materials to avoid excessively long output files; otherwise, eliminate this printout by using

```
Print -85 -86
```

Figure 2.21 is an MCNP plot of basic three electron materials. The MCNP code is run in the cross-section plotting mode:

```
MCNP6  I=Ex12_2c  name=junk.  IXZ
```

And the MCNP plot command is

```
ylim 1 200 par e xs m609 cop xs m201 cop xs m313
```

The total, collisional, and radiative stopping powers are plotted in Fig. 2.22 from the same input file with the following command:

```
par e xs m313 ylim .5 30 mt 3 cop mt 1 cop mt 2
```

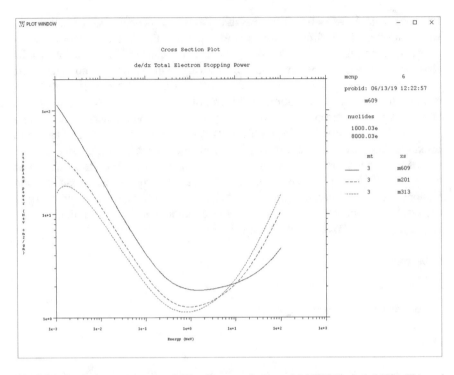

Fig. 2.21 Total electron-stopping (MT = 3) power for material M609 (water), M201 (Zr), and M313 (UO$_2$)

The stopping powers are for materials but not nuclides. Thus, in the cross section, plot command XS = M313 works, but XS = 92000.03e does not. Electron-stopping powers also depend on the material density. Thus, if a material is at more than one density, the stopping powers for the first density will be correct but wrong for all others. Each material of a different density requires a new material number even though the nuclides are the same.

2.2.7 Data and Models for Ions and Charged Particles

The MCNP code models 36 light ions and subatomic particles and also transports 2200+ heavy ions. See "MCNP6 Particles" Table 2-2 in the MCNP code manual. Though there are a few data libraries for protons, deuterons, tritons, alphas, and helions, most nuclides and all the other particle types (deuteron o, triton r, alpha a, helion s) and heavy ions use physics models and continuous slowing-down theory (like electrons). A number of physics models have been developed internationally and these can be selected along with key parameters on the LCA, LCB, LEA, LEB,

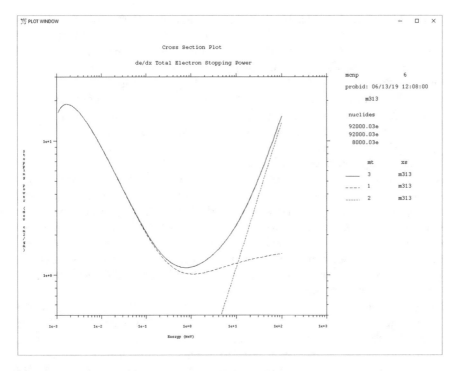

Fig. 2.22 Electron-stopping powers for material M313, UO$_2$: total (MT = 3), collisional (MT = 1) and radiative (MT = 2)

and LCC input cards. The principal MCNP models are CEM, LAQGSM (default) and Bertini, Isabel, and INCL, which have their own specific data files (such as BERTIN, FALPHA, and CINDERGL) provided in the MCNP code package. See the MCNP manual [2] section 3.3.3.7.1.

The following example shows how the MODE, M0, and M111 cards might be changed for a charged particle problem:

```
MODE    N  P  E  H  D  T  S  A  |  /  Z
m0 nlib=.70c plib=.04c elib=.03e alib=.00a slib=.00s
MT111 POLY BENZ GRPH
M111    6000     1 1001 1     7014 1
MX111:P 6012.24U   0           7014.70U  $ Photonuclear
MX111:H 6012.70H   1001.71H    7014.24H  $ Proton
MX111:D 6012.00o   1001.00o    7014.00o  $ Deuteron
MX111:T 6012.00R   1001.00R    7014.00R  $ Triton
MX111:S 6012.00S   MODEL       7014.00S  $ Helion
MX111:A 6012.00A   MODEL       8016.00A  $ Alpha
```

The MODE card will enable transport of neutrons (N), photons (P), electrons (E), protons (H for hadron), deuterons (d), tritons (T), helions (S for ³He), and alphas (A). Good practice is to add negative muons (|), positive pions (/), and neutral pions (Z) to the MODE card when energies in the problem reach the level of rest mass for these particles (~100–150 MeV). Add positive kaons (K) for energies >~500 MeV. Electron neutrinos (U) and muon neutrinos (V) often are added.

The M0 card is the default for all materials cards. NLIB, PLIB, ELIB, ALIB, and SLIB are for neutron, photoatomic, electron, alpha, and helion data.

The MX cards will enable nuclide substitution. No substitutions are allowed for photoatomic (P) and electron (E) data because these data are elemental rather than isotopic. Carbon on the M111 card is specified as 6000 for all modern carbon cross-section libraries, but carbon is available only as 6012 for photonuclear, proton, and other light nuclei libraries, thus requiring the MX111 cards. Also, isotopes rather than elements should be specified for all model physics. MX111:P is for photonuclear data and specifies that no cross section is used for photonuclear hydrogen because hydrogen has no photonuclear reactions. The library types for deuterons, tritons, helions, and alphas are O, R, S, and A. MX111:S and MX111:A are for helions and alphas and specify model physics rather than data libraries, which are used for hydrogen. For neutrons, photoatomic, photonuclear, and electrons, the data libraries are best in the energy range for which they are provided. For protons, the data libraries—if they are available—are sometimes better for secondary particle production but not transport, where models are sometimes better. For the remaining charged particles and ions, the models are generally better; data libraries, such as TENDL (https://tendl.web.psi.ch/tendl_2019/tendl2019.html) must be used with caution.

2.2.8 Additional Data Diagnostics and Recommendations

As illustrated in the previous sections, MCNP data can be plotted. See Sect. 2.5 for details.

The MCNP code has a "first collision" feature for studying the physics in isolation. It is activated by the 8th entry on the LCA card (MCNP manual section 3.3.3.7.2):

```
LCA  7J  -2.
```

Source particles do not transport and instead initiate an immediate collision. Products of the reaction are then transported out as if in a void. The option works for library and model physics. With this option, F1 or other tallies can be used to tally angle dependence, energy dependence, multiplicities, and secondary particle production of any particle type that collides with any nucleus.

The MCNP code also has a GENXS option. With a beam of particles or nuclei impinging upon a thin target and the GENXS option, the MCNP code can easily calculate double-differential and angle/energy integrated spectra, multiplicities, yields of emitted particles, fission cross sections, and fission yields. The GENXS option is like using the event-generator of the MCNP6 code as a stand-alone code—without any real transport through the matter. A small and simple second auxiliary input file is required and is described along with the GENXS option in the MCNP manual (section 3.3.3.9).

The following recommendations avoid common user errors:

- Mn material card charged particle steps:

  - ESTEP (electrons) and hstep (all other charged particles) have a default of 3 substeps, which may be too small; try something larger like 10 or 20 to see if results—particularly secondary energy distributions—change, although this will run slower;
  - EFAC on the PHYS card for each charged particle type controls the slowing-down energy step size and the default, EFAC = 0.917, may be too small; try something larger like 0.99 to see if results change, although this will run slower;

- Upper energy bounds: The PHYS:x for all particle types has E_{max} as the first entry, which must be set above maximum energy of the problem to prevent wrong answers. It is recommended to set it just above to avoid inefficient effects of loading unneeded data;
- MODE n h d t s a l / z k --- add d t s a as needed; l / z for $E > 100$ MeV; k for $E > 500$ MeV.
- Lower energy cutoff. On the CUT:x card, set the 2nd entry, Emin, to 0.001 to go below the high default E lower energy cutoff. The default cutoff is the A number of the projectile. Thus, alpha particles transported below 4 MeV will be absorbed unless E_{min} is lowered.
- Consider analog capture, 3rd and 4th entry = 0, on CUT:x, to avoid 0 weight window complications.

2.3 Sources

2.3.1 SDEF Fixed Sources

The input in Example 2.12 is similar to Example 2.6 with additional materials specifications. It is illustrated in Fig. 2.14 through Fig. 2.16.

Example 2.12 Fully Specified FILL Square and Hex Lattices. Square and Hex Lattices

```
1   313  -10.44      -1        u=1    imp:n=1   $ UO2
2   201   -6.55      -2 1      u=1    imp:n=1   $ Zr clad
3   609   -1.00       2        u=1    imp:n=1   $ H2O
```

```
21 111  -4.00    -31 lat=1 u=11  imp:n=1  $ Square lattice
            fill = -3:3 0:0 -3:3
                        1 1 1 1 1 1 1
                        1 1 1 1 1 1 1
                        1 1 1 1 1 1 1
                        1 1 1 1 1 1 1
                        1 1 1 1 1 1 1
                        1 1 1 1 1 1 1
                       11 1 1 1 1 1 1
22 111  -4.00    -32 lat=2 u=12  imp:n=1  $ Hexagonal lattice
          fill =  -5:5 -4:4 0:0
                      0 0 0 0 1 1 1 1 1 1 1
                       0 0 0 1 1 1 1 1 1 1 1
                        0 0 0 1 1 1 1 1 1 1 0
                         0 0 1 1 1 1 1 1 1 1 0
                          0 0 1 1 1 1 1 1 1 0 0
                           0 1 1 1 1 1 1 1 1 0 0
                            0 1 1 1 1 1 1 1 0 0 0
                             1 1 1 1 1 1 1 1 0 0 0
                            12 1 1 1 1 1 1 0 0 0 0
14   0    -4              fill  11 imp:n=1  $ Square lattice box
11 like 14 but TRCL=(20 0 0) fill 12       $ Hexagonal lattice box
15   0  #14 #11 -20             imp:n=1  $ outside boxes
16   0            20            imp:n=0  $ outside world

1   rcc  0 0 0   0 12 0   .5      $ Right circular cylinder
2   rcc  0 0 0   0 12 0   .6      $ Right circular cylinder
4   rpp -7 7    0 12  -7 7      $ Right parallelepiped 14x12x14
31  rpp -1 1    0 12  -1 1      $ Right parallelepiped 2x12x2
32  rhp  0 0 0   0 12 0  1 0 0  $ Right hex 12x1
20  sph 0 0 0 50                $ Sphere containing lattice boxes

MODE N P
m0 nlib=.70c plib=.04c elib=.03e
m201   40090 -0.505239    40091 -0.110180    40092 -0.168413
       40094 -0.170672    40096 -0.027496    nlib=80c
m313   92235 -0.02759   92238 -0.85391    8016 -0.11850
M609   1001 2   8016 1
MT609  LWTR.10t
M111   6000 1   1001 1   7014 1
MT111 POLY.10t BENZ.10t GRPH.10t
sdef   pos=10 6 0
nps   1000000
PRINT
```

The MCNP standard source definition, SDEF, may have no entries, fixed entries, distributions, and dependent distributions.

The SDEF in Example 2.12 places the source point at position POS=10 6 0 instead of the default POS=0 0 0 cm

```
sdef  pos=10 6 0
```

Example 2.12 SDEF is the same as

```
SDEF  X=10 Y=6 Z=0  ERG=14  TME=0  PAR=N  CEL=15 EFF=.01 WGT=1.0
```

Dimensions (POS, X, Y, Z, RAD, EXT) are in cm, energies (ERG) in MeV, and times (TME) in shakes, 10^{-8} s.

Particle types are listed in Table 2-2 of the MCNP manual. In addition, heavy ions may be source particles, as in PAR = 6012, which requires heavy ions to be specified on the MODE card as MODE #. Antiparticles are specified as negative: PAR = −E is a positron source. Particle types may also be PAR = SF spontaneous fissions, PAR = −SF spontaneous fission neutrons, or PAR = SP spontaneous photons.

The default cell is wherever the source is located. If the cell is specified, CEL=15, then any part of the source not in that cell is rejected. Consequently, CEL specifies a rejection criterion rather than a cell location. If more than 1% of the source locations are rejected, then the calculation is stopped because the default source rejection efficiency is EFF=0.01. In some calculations, it is desirable to set the source rejection efficiency lower, such as EFF=0.0001.

The particle weight, WGT, is the physical number of particles modeled in the source. The default is 1 source particle so that all results may be easily normalized to the true source strength. All results are normalized to the source weight, WGT.

Additional SDEF scalar variables are SUR, NRM, ARA, TR, and CCC (see table 3-62 of the MCNP manual).

Particles starting exactly on a surface must have that surface name specified with SUR. If the particles are directed inward or outward relative to the source normal, then NRM is specified as NRM=1 in the direction of the positive source normal and NRM=−1 in the direction of the negative source normal.

If a source is mono-directional, then the surface area of the source, ARA, must be specified if there are point detector tallies, F5, or DXTRAN variance reduction.

The location of source particles may also be translated and rotated, where TR=n refers to the TRn card translation and rotation.

Source particles may also be rejected if they are outside a region defined by a cookie cutter cell (CCC), which is not necessarily a cell in the problem.

2.3.2 SDEF Source Distributions

Additional SDEF keywords are vectors VEC and AXS, which are associated with distributions RAD and EXT.

In the following example, only the region inside the hexagonal box is sampled after starting particles uniformly in volume in an overlaying sphere. Replace the SDEF in Example 2.12 with

```
SDEF    POS=10 6 0    RAD=D11   CCC=11   EFF=.0001
SI11    0 20
SP11   -21 2
```

The combination of POS and RAD samples sources particle location in a sphere. Radius is now a distribution, RAD=D11. The source information card, SI11, distributes the radius between 0 and 20 cm. The source probability, SP11, invokes a power law (distribution type −21), where the radius is raised to the second power in order to sample uniformly in volume of the sphere. The second power is needed because the differential volume of a sphere is

$$dV = 4\pi r^2 dr$$

Note that CCC rejects any point sampled in the sphere volume that is not in the cookie cutter cell, CCC=11, which is the hexagonal lattice box. The efficiency of sampling is 0.0696. Whereas some source particles do not find a source location within the box after 100 tries, MCNP stops with the error message

```
expire parameter is the sampling efficiency in source cell 11 is too low
```

The problem immediately terminates with a "bad trouble" error; therefore, the efficiency must be set lower. In this case, EFF=0.0001 works.

To sample a cylindrical volume, the combination POS, RAD, EXT is used (replacing SDEF in Example 2.12):

```
SDEF    POS=0 6 0    RAD=D12    EXT=D13    AXS=0 1 0
SI12    0 10
SP12   -21 1
SI13 H -6 6
SP13    0 1
```

The radius is sampled (SI12) between 0 and 10 cm as a power law raised to the 1st power (SP12 −21 1) because the differential volume of a cylinder with respect to r is

$$dV = 2\pi rh \, dr$$

The extent is sampled in the Y-direction (AXS=0 1 0) as a histogram (SI13 H −6 6) from −6 < Y < 6 cm relative to the position (POS=0 6 0). The distribution is linear with probability (SP13 0 1) 0 for Y < −6, 1 between −6 < Y < 6. The H designator is one of the several possibilities:

H—histogram bin <u>upper</u> boundaries *(default)*
L—discrete values follow
A—points where probability density distribution is defined
S—distribution numbers follow

The SP options are as follows:

D—bin probabilities for an H or L distribution *(default)*
C—cumulative bin probabilities for an H or L distribution
V—probability is proportional to cell volume
W—intensities for a mix of particle sources

Particles may be started only in cell 1 of the square lattice (replacing SDEF in Example 2.12) by making CEL a distribution; however, the full path through the universe levels must be specified when the cell is in a universe. Particles are started in all 48 locations of cell 1:

```
SDEF   POS=0 6 0   RAD=D12   EXT=D13   AXS=0 1 0   CEL=D14   EFF=1e-5
SI12   0 10
SP12   -21 1
SI13 H -6 6
SP13 D  0 1
SI14 L (1<21<14)
SP14    1
```

Note that EFF had to be set because the efficiency is 0.0025, less than the default 0.01.

Particles may also be started in all 66 locations of cell 1 of the hexagonal lattice by specifying which filling universes of cell 22 to use. Because the universe ($u = 1$) is specified, the probabilities of the individual cell 1s must be specified (SP14) and can therefore be different if desired:

```
SDEF   POS=0 6 0   RAD=D12   EXT=D13   AXS=0 1 0   CEL=D14
SI12   0 .5
SP12   -21 1
SI13   -6 6
SP13   0 1
SI14 L (1<22[u=1]<11)
SP14   1 65R
```

In both this and the previous case, the radius may simply be the cell 1 radius (SI12 0 .5) and no EFF is needed because the source efficiency is 100% in each cell 1.

The most common specification of cells in a lattice specifies the path to each individual repetition of the cell. In the following, particles are started in all 66 locations of cell 1 of the hexagonal lattice but with different probabilities depending upon whether the cell is cut off or not (Fig. 2.14):

```
SDEF   POS=0  6  0   RAD=D12   EXT=D13  AXS=0  1  0   CEL=D14
SI12    0 .5
SP12   -21  1
SI13 H -6  6
SP13    0  1
SI14   L  (1<22[-5      3   0]<11)
          (1<22[-4  1:4   0]<11)  (1<22[  -3 -1:4   0]<11)
          (1<22[-2 -3:4   0]<11)  (1<22[-1:1 -4:4   0]<11)
          (1<22[ 2 -4:3   0]<11)  (1<22[   3 -4:1   0]<11)
          (1<22[ 4 -4:-1 0]<11)  (1<22[   5 -4:-3 0]<11)
SP14       .5
           .5 1 1 .5                    .5 1 3R .5
           .5 1 5R .5 .5 1 6R .5 .5 1 6R .5 .5 1 6R .5
           .5 1 5R .5                   .5 1 3R .5
           .5 1 1 .5                    .5 .5
```

The square brackets indicate the lattice index numbers. The ":" notation on the SI entries expands the 9 entries to the 66 lattice elements. The SP entries are .5 for elements that are cut off by the filling cell in the universe level above.

2.3.3   SDEF Dependent Distributions: DS

The SDEF definition and source distributions from Example 2.19 in Sect. 2.4.3 and shown in Example 2.13 illustrate dependent distributions. The MCNP code allows one level of dependency, but this example illustrates how to model three levels of dependency: energy is dependent upon time which is dependent upon location dependent upon the particle type.

Example 2.13 Dependent Source Description Taken from Example 2.19

```
SDEF   PAR=D42
       CEL=FPAR=D51   POS=FPAR=D53   RAD=FPAR=D54   EXT=FPAR=D55
       AXS=FPAR=D56   TME=FPAR=D57   ERG=FPAR=D58
SI42   L  SF   N   P    P    P    P                            $ PAR
SP42      .3  .3  .1   .1   .1   .1                            $ PAR
DS51   S  46  47  48   48   48   48                            $ CEL
```

```
DS53    L   0  6  0    0  6  0    10  3  0    10  3  0    10  7  0    10  7  0        $ POS
DS54    S   43   43    0    0    0    0                                              $ RAD
DS55    S   44   44    0    0    0    0                                              $ EXT
DS56    S   45   45    0    0    0    0                                              $ AXS
DS57    L   0  1e8  2e8  3e8  4e8  5e8                                               $ TME
DS58    S   0  92  93  94  95  96                                                    $ ERG
SC46    Spontaneous fission cells in XYZ lattice
SI46    L  (1<21[-2:3  0  3]<14)  (1<21[-3:3  0:0  -3:2]<14)
SP46       1  5R                          1  41R
SC47    Neutron source cells in HEX lattice
SI47    L  (1<22[-5      3   0]<11)
           (1<22[-4   1:4   0]<11)  (1<22[   -3  -1:4   0]<11)
           (1<22[-2  -3:4   0]<11)  (1<22[-1:1  -4:4   0]<11)
           (1<22[ 2  -4:3   0]<11)  (1<22[    3  -4:1   0]<11)
           (1<22[ 4  -4:-1  0]<11)  (1<22[    5  -4:-3  0]<11)
SP47       .5
           .5 1 1 .5                       .5 1 3R .5
           .5 1 5R .5 .5 1 6R .5 .5 1 6R .5 .5 1 6R .5
           .5 1 5R .5                    .5 1 3R .5
           .5 1 1 .5                     .5 .5
SC48    Point photon source cell in water between lattices
SI48    L  15
SP48       1
c       particle-dependent radius, extent, axis
SI43     0 .5                                                                       $ Radii
SP43    -21  1                                                                       $ Radii
SI44     -6  6                                                                       $ Extent
SP44    -21  0                                                                       $ Extent
SI45    L   0 1 0                                                                    $ Axis
SP45       1                                                                         $ Axis
c       energies dependent upon time, position, and particle type
SP92    -3 1.175 1.0401              $ 252Cf Frohner Watt parameters
SI93    50 59  $ Photon, XYZ water, TME 2e8, energies
SP93     0  1
SI94    60 69  $ Photon, XYZ water, TME 3e8, energies
SP94     0  1
SI95    L 70  $ Photon, XYZ water, TME 4e8, energies
SP95       1
SI96    L 80  $ Photon, XYZ water, TME 5e8, energies
SP96       1
```

The SDEF definition has particle type (PAR) as an independent distribution and then dependent distributions for cell (CEL), position (POS), radius (RAD), extent (EXT), axis (AXS), time (TME), and energy (ERG).

The particle types are spontaneous fission (SF), neutrons (N), and then photons (P). The photons are repeated four times (SI42) to enable more levels of dependencies: there is an energy distribution for each time, times are sampled from two locations, and locations are sampled from three particle types.

The DS cards point to the values or distributions for each of the six particle types.

DS51 provides the distributions of cells for each particle type. Spontaneous fission occurs in the XYZ lattice cells, distribution 46. Note that each spontaneous fission will produce between zero and ten spontaneous fission neutrons. The neutron source is in the HEX lattice cells, distribution 47. The photon source is in cell 15, distribution 48. Note that each of these distributions has a title description on the source comment (SC) card.

DS53 provides the source position for each particle type. Note that the position in the XYZ lattice and the HEX lattice is the same, namely the [0 0 0] lattice element is a POS=0 6 0 in the coordinate system of the filling square lattice boxes, which are 20 cm apart. The photons start between the lattice boxes at POS=10 3 0 or POS=10 7 0.

DS54, DS55, and DS56 point to the distributions for radius, extent, and axis for the fuel pin cylinders. Whereas the photon source is at points, there is no radius, extent, or axis description, so the photon distributions are set to distribution zero or no distribution.

DS57 sets the time of spontaneous fission (0 s), neutron source (1 s), and four photon times, 2, 3 4 and 5 s at POS=10 3 0 and 4 and 5 s at POS=10 7 0.

DS58 specifies the energy distributions. None is specified for spontaneous fission because the appropriate Watt spectrum for each spontaneous fission is automatically sampled by the MCNP code. Neutrons are started in the HEX lattice with a ^{252}Cf Frohner Watt spectrum [5], SP92. Note that no SI92 is used for special Watt distribution (-3) on SP92. The photon energies are

$50 < E < 59$ MeV for TME=2 s and POS=10 3 0;
$60 < E < 69$ MeV for TME=3 s and POS=10 3 0;
$E = 70$ MeV for TME=4 s and POS=10 7 0; and
$E = 80$ MeV for TME=5 s and POS=10 7 0

Less common options for the dependent distribution DS card are H for dependent continuous histogram and T and Q for dependent values of independent values. For example,

```
ERG = D1      DIR = FERG = D3
SI1   .001   .01   .1   .4   1.3   2.6   5   14
DS3 Q .3   4   1.8   5   14   6
```

The DS3 card with the Q option specifies

From 0–0.3 MeV	Get DIR from distribution SI4
From 0.3–1.8 MeV	Get DIR from distribution SI5
From 1.8–14 MeV	Get DIR from distribution SI6

2.3.4 Criticality Sources

MCNP criticality calculations are used in safeguards to ensure criticality safety. Nuclear criticality is the determination of k_{eff}, which is the number of neutrons in a system produced by an initial number of neutrons. If $k_{eff} < 1$, the system is subcritical, and more neutrons leak out or are captured than those produced by fission. If $k_{eff} = 1$, then the system is critical, and the number of neutrons produced by fission precisely balances the number lost to leakage and other physical processes [6]. See Example 2.14.

Example 2.14 Four 20 wt% Enriched Uranium Cans (Radius 8 cm, Height 40 cm) in Air (Radius 25 cm, Height 42 cm) Inside an Aluminum Casing (Radius 40 cm, Height 44 cm) with Criticality (KCODE) Source

```
Four uranium cans in air and aluminum
11 101 -18.0     -51                  imp:n=1 $ uranium can
12 like 11 but trcl=(20  0 0)         imp:n=1 $ uranium can
13 like 11 but trcl=( 0 20 0)         imp:n=1 $ uranium can
14 like 11 but trcl=(20 20 0)         imp:n=1 $ uranium can
21 102  -1.0  -52 #11 #12 #13 #14     imp:n=1 $ air tank
22 103  -2.7  -53 52                  imp:n=1 $ Al liner
23  0              53                 imp:n=0 $ outside

51 RCC  0  0 -20  0 0 40   8                   $ fuel cylinder
52 RCC 10 10 -21  0 0 42   25                  $ air cylinder
53 RCC 10 10 -22  0 0 44   40                  $ aluminum liner

m101  92238 -.8   92235 -.2 nlib=.80c
m102  7014.80c .8   8016.80c .2
m103  13027.80c 1
print
kcode 1e4 1.0 30 130
sdef  axs=0 0 1 rad=d31 ext=d32 pos=d33
si31  0 8
sp31 -21 1
si32 -20 20
sp32   0 1
si33  L 0 0 0  20 0 0  0 20 0  20 20 0
sp33  1 3r
FC1   Neutrons crossing surfaces
F1:n (51.1 51.2 51.3) (52.1 52.2 52.3) (53.1 53.2 53.3)
C1    0 1 T
FQ1   F C
FC4  Fissions and neutron production in uranium cells
```

```
F4:n   11 12 13 14 T
FM4  (-1 101 -6)  (-1 101 -6 -7)
FQ4    F  M
SD4    1 4r
```

Figures 2.23 and 2.24 show the axial and cut-away views of the criticality problem of Example 2.14.

Example 2.14 is a contrived model of four uranium cans in air surrounded by aluminum. The source is specified by a KCODE card

```
KCODE     1000    1.0    30     130
```

The KCODE default values are shown above.

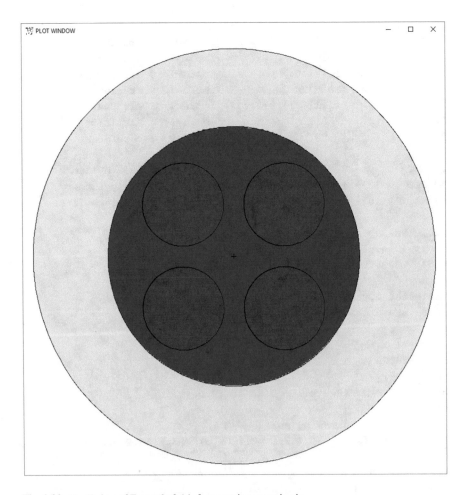

Fig. 2.23 $X - Y$ view of Example 2.14, four uranium cans in air

The first KCODE entry is the number of histories to run in a generation, or "batch" or "cycle." This number should be as large as possible in order to sample all geometry, particularly fissionable regions. Because each generation of fission neutrons is based on the previous generation, the statistical sampling of k_{eff} is correlated and thus violates the central limit theorem of statistics. Consequently, the estimated standard deviation of k_{eff} is underpredicted, which can yield incorrect results. Further, nuclear systems must be "delayed critical," which means that the part of k_{eff} from delayed neutrons—less than 1% of the fission neutrons produced—requires that the relative error allowed for k_{eff} is $\ll 0.01$. Further, variance reduction cannot be used to better converge results because k_{eff} is based on the balance of the whole system and variance reduction works by focusing on important parts of phase space at the expense of others.

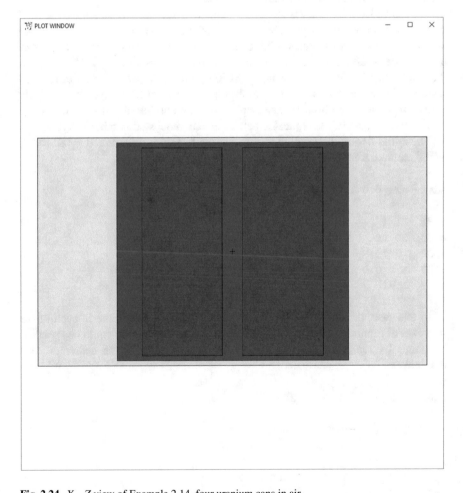

Fig. 2.24 $X - Z$ view of Example 2.14, four uranium cans in air

The second KCODE entry is the guess of k_{eff} for the first cycle only. It does not need to be predicted closely and can be a factor of two off. If it is too low, then MCNP will issue the message

```
new source has overrun old source.  will start over with larger k.
```

If the guess is too high, then MCNP will waste a lot of storage space. So the default of 1.0 is fine.

The third KCODE entry is the number of settle cycles to run, that is, the number of cycles to be skipped before beginning tally accumulation. Whereas the initial fission distribution must be specified by KSRC, SDEF, or SRCTP, a number of cycles are needed to better distribute the fission source before beginning the calculation of k_{eff} and any tallies. The default of 30 should be the minimum. If k_{eff} or tallies of interest do not converge, then more may be needed.

The fourth KCODE entry is the total number of cycles to run altogether. The total number of histories in the problem is approximately the product of the batch size and the total number of cycles. With these default values, the first 30,000 histories are discarded—after hopefully converging the fission distribution—and then k_{eff} and the tallies are calculated from the next 100,000 histories, for a total of about 130,000 histories. Note that the batch size varies slightly from cycle to cycle to correctly adjust for particle weighting; consequently, the final number of histories will not be exactly the product of the requested batch size and requested number of cycles.

The initial fission distribution for the first cycle only is determined by KSRC or SDEF cards or a SRCTP file. Example 2.14 uses

```
sdef  axs=0 0 1 rad=d31 ext=d32 pos=d33
si31   0 8
sp31  -21 1
si32  -20 20
sp32   0 1
si33  L 0 0 0  20 0 0  0 20 0  20 20 0
sp33   1 3r
```

In this case, the initial fission distribution is uniform in volume in each of the four cans of uranium. Alternatively, the initial fission distribution could be specified by points by replacing the above with one fission event in the center of each uranium can

```
KSRC   0 0 0  20 0 0  0 20 0  20 20 0
```

which is the same as

```
sdef  pos=d33
si33  L 0 0 0  20 0 0  0 20 0  20 20 0
sp33   1 3r
```

Alternatively, and best, is to use the SRCTP file from one run as the initial source for the next if all the fission cells have the same geometry. The SRCTP has the source locations and energies from a previous calculation which is better than an approximate guess. In this case, the name of the source file is specified on the MCNP execution line and neither KSRC nor SDEF is present.

Example 2.14 also has two tallies specified after the source description. These are described in more detail in Sect. 2.4, Output and Tallies, and are the number of neutrons crossing surfaces 51, 52, and 53 and the number of fissions and number of fission neutrons in cells 11, 12, 13, and 14.

Because k_{eff} must be known very precisely, with an error $\ll 1\%$, and because the relative errors of tallies and standard deviations of k_{eff} are underestimated by violating the central limit theorem of statistics, the MCNP code calculates thousands of values of k_{eff} in each run. The MCNP code calculates k_{eff} in seven different ways using three uncorrelated estimators (collision, absorption, track length) and four additional combinations (collision/absorption, absorption/track length, track length/collision, collision/absorption/track length). These estimators and combinations are for every cycle individually, as simple averages and as covariance-weighted averages. Fortunately, the "the final estimated combined collision/absorption/track-length k_{eff}" is printed in a box in the output file. How well this final k_{eff} is converged is indicated by how well the k_{eff} values agree and deviate from the final.

We highly recommend tallies for KCODE calculations to provide confidence in the overall calculation of k_{eff}. Tally 4 in Example 2.14 uses the FM card (see Sect. 2.4.3.6) to multiply the flux in each fissionable can by $\rho\sigma_f$ and $\rho\nu\sigma_f$ to get the number of fissions and number of fission neutrons produced. The tally 4 output is

cell	number of fissions		number of fission neutrons	
11	7.77321E-02	0.0018	1.99838E-01	0.0018
12	7.68256E-02	0.0018	1.97500E-01	0.0018
13	7.74996E-02	0.0018	1.99243E-01	0.0018
14	7.79120E-02	0.0018	2.00343E-01	0.0018
total	3.09969E-01	0.0007	7.96924E-01	0.0007

The total number of fission neutrons agrees well with "the final estimated combined collision/absorption/track-length $k_{eff} = 0.79691$ with an estimated standard deviation of 0.00056"; however, the tallies between the different fissionable cells should be the same due to symmetry. Instead, they are 1.44% apart with 0.0018 relative errors, which implies that the fission distribution is not converged. We would recommend running ten times as many histories per cycle, KCODE 1e5, to seek better convergence of the fission distribution. But since $k_{eff} \ll 1.0$, further accuracy may not be needed because the system is subcritical below any reasonable margin of error.

Section 2.5 describes how to plot tally output. Figure 2.34 in that section is an MCNP tally plot showing the convergence of this Example 2.14 criticality calculation. The combined C-A-T—collision, absorption, and track length—tallies are plotted together as a function of cycle number and the plot commands are provided.

2.3.5   *Surface Source Write and Read (SSW, SSR)*

The surface source enables particles that cross various surfaces in a problem to be
recorded (SSW—surface source write run) and then rerun in a subsequent problem
(SSR—surface source read run), omitting all surfaces leading up to it. Thus, the
subsequent SSR run can be used for multiple parameter studies without having to
repeat the tracking through the initial part of the problem.

   A frequent use in safeguards is to model a bremsstrahlung source with full and
time-consuming electron transport in the first SSW run, writing the resulting photons to a surface source file. In the subsequent SSR run, only the photons from the
surface source file are followed without the expensive initial electron run. See
Example 2.15.

**Example 2.15 SSW Input File to Write a Surface Source. The Geometry is the
Same as That for the Criticality Example 2.14: Four 20 wt % Enriched Uranium
Cans (Radius 8 cm, Height 40 cm) in Air (Radius 25 cm, Height 42 cm) Inside
an Aluminum Casing (Radius 40 cm, Height 44 cm) with a Fixed SDEF Source**

```
Four uranium cans in air and aluminum
11 101 -18.0      -51                    imp:n=1  $ uranium can
12 like 11 but trcl=(20   0 0)           imp:n=1  $ uranium can
13 like 11 but trcl=( 0 20 0)            imp:n=1  $ uranium can
14 like 11 but trcl=(20 20 0)            imp:n=1  $ uranium can
21 102    -1.0   -52 #11 #12 #13 #14     imp:n=1  $ air tank
22 103    -2.7   -53 52                   imp:n=1  $ Al liner
23    0                  53               imp:n=0  $ outside

51 RCC  0   0 -20  0 0 40    8                    $ fuel cylinder
52 RCC 10 10 -21  0 0 42    25                    $ air cylinder
53 RCC 10 10 -22  0 0 44    40                    $ aluminum liner

m101   92238 -.8   92235 -.2 nlib=.80c
m102   7014.80c .8   8016.80c .2
m103   13027.80c 1
print
nps    100000
sdef   axs=0 0 1 rad=d31 ext=d32 pos=d33
       erg=d34
si31   0 8
sp31  -21 1
si32  -20 20
sp32    0 1
si33   L 0 0 0   20 0 0   0 20 0   20 20 0
sp33   1 3r
```

```
sp34   -3 .965 2.29
FC1:n Neutrons crossing surfaces
F1:n  (51.1 51.2 51.3)  (52.1 52.2 52.3)  (53.1 53.2 53.3)
C1      0 1 T
FQ1     F C
FC4   Fissions and neutron production in uranium cells
F4:n   11 12 13 14 T
FM4  (-1 101 -6) (-1 101 -6 -7)
FQ4     F M
SD4    1 4r
SSW       52.1      52.2      52.3
```

Example 2.15 is the same geometry as the criticality Example 2.14 illustrated in Figs. 2.23 and 2.24. The geometry of Example 1.3-3 is used here to avoid introducing an entirely new geometry. But the criticality source (KCODE) is replaced with surface source write (SSW) and surface source read (SSR) making Example 2.15 an entirely different problem for illustrative purposes. The criticality source

```
kcode 1e4 1.0 30 130
sdef   axs=0 0 1 rad=d31 ext=d32 pos=d33
```

is replaced with

```
nps    100000
sdef   axs=0 0 1 rad=d31 ext=d32 pos=d33
       erg=d34
sp34   -3 .965 2.29
```

where the energy distribution, SP34, is a Watt spectrum (special function −3) with the Watt parameters of the initial default KCODE cycle fission distribution ($a = 0.965$ MeV, $b = 2.29$ MeV^{-1}).

In addition, the surface source write card is added

```
SSW       52.1      52.2      52.3
```

which writes all particles crossing surface 52 between the air and the aluminum liner to a surface source write file, WSSA. Note that the surface must be specified as facets and that surfaces created with LIKE BUT, such as 12051.1, may not be used. Note that the name option will write a file with .w appended

```
MCNP6    I=inputfile    name=outSSW.
```

will rename the WSSA file as outSSW.w

The resulting tally for surfaces 51, 52, and 53 is

```
1tally   1      nps = 100000         Neutrons crossing surfaces
       surface  a is (51.1 51.2 51.3)
       surface  b is (52.1 52.2 52.3)
       surface  c is (53.1 53.2 53.3)

 angle   :   0.0000E+00            1.0000E+00                  total
surface
     a  9.72549E-01 0.0079  1.99828E+00 0.0071  2.97083E+00 0.0073
     b  1.29713E+00 0.0071  3.72160E+00 0.0062  5.01873E+00 0.0064
     c  0.00000E+00 0.0000  2.41904E+00 0.0060  2.41904E+00 0.0060
```

The SSR run input file that utilizes the resulting surface source file written by the SSW run is shown in Example 2.16.

Example 2.16 SSR Run Input File Reading the Surface Source File Produced by Example 2.15

```
Air tank and aluminum casing
21 102  -1.0   -52                    imp:n=0  $ air tank
22 103  -2.7   -53 52                 imp:n=1  $ Al liner
23   0          53                    imp:n=0  $ outside

52 RCC 10 10 -21   0 0 42   25                 $ air cylinder
53 RCC 10 10 -22   0 0 44   40                 $ aluminum liner

m102   7014.80c .8   8016.80c .2
m103   13027.80c 1
print
nps    200000
SSR   OLD = 52.1 52.2   52.3
FC1:n Neutrons crossing surfaces
F1:n (52.1 52.2 52.3) (53.1 53.2 53.3)
C1     0 1 T
FQ1    F C
```

Example 2.16 is the input file that utilizes the surface source produced by Example 2.14. It is run with an execution line that names the surface source file to be used.

```
MCNP6   I=exSSR   name=outSSR.   RSSA=outSSW.w
```

Note that the interior of the air tank is deleted, along with its surfaces and materials. The importance of the air (cell 21) is set to zero so that any neutron backscattering out of the aluminum liner is killed. This important albedo consideration is because all particles scattering from the liner back into the air were accounted for already in the surface source and must not be counted twice. The surface source read card

```
SSR   OLD = 52.1 52.2   52.3
```

specifies the surfaces to be read. They may also have new names in the subsequent SSR run as long as they have the same geometric location.

The full SSR card is

```
SSR   old=S₁ S₂   new=S₃ S₄   cel=C₁ C₂   col=n WGT=w TR=m PSC=x
```

where old and new surface names are given; for example,

```
SSR   OLD = 52.1 52.2   NEW = 62.1 62.2    TR=123
```

reads the tracks crossing surfaces 52.1 and 52.2, which are started on identical surfaces 62.1 and 62.2 located elsewhere according to surface transformation TR123 (not shown). Surface 52.3 is not read.

For both the SSW and SSR, cel=C_1 C_2 is used to record all fission sites in a KCODE calculation. Additional options are available on the SSR card (MCNP manual section 3.3.4.8). The entry col=n filters $n = -1/0/1$ uncollided, all, collided particles. WGTcan renormalizes particle weights. TR maps the surface elsewhere, or it can be a distribution to repeat it many places in the subsequent run. PSC is used only for point detector (F5) tallies and DXTRAN variance reduction.

The surface source that is written in the initial run is described in the output file of the run that reads it. For Example 2.16, the SSR output file describes the surface source it is reading

```
1summary information from this use of surface-source file outSSW.w
print table 170


     number of original histories       100000
     number of records read             677171
     number of recorded histories        99271
     number of histories read            99271


     source multiplication factor      2.00000

     total source weight                          3.72160E+05
     total rejected weight                        0.00000E+00
     fraction weight rejected                     0.00000E+00
```

In Example 2.15, 100,000 histories were run, causing many fissions to produce 677,171 tracks crossing surface 52. The number of recorded histories is 99,271, indicating that 729 histories and the resulting fission progeny did not cross surface 52. The source multiplication factor of 2.0 is because NPS is 100,000 in the SSW Example 2.15 and 200,000 in the SSR Example 2.16. If NPS is different in the SSW writing and SSR reading runs, the tracks crossing the surface will be split or rouletted accordingly.

Particles or tracks written to the surface source file may be rejected if they are for a particle not listed on the subsequent run MODE specification, if they are filtered by col=n, if they are not on one of the listed OLD surfaces, or if they are outside the time or energy cutoffs of the subsequent problem.

The output of Example 2.16 summarizes the surface source read as

```
total number of particles accepted                  677171
total weight accepted                             3.72160E+05
fraction of total weight accepted                 0.10000E+01
```

That is, 677171 tracks crossed surface 52, with a total weight of 3.72160E5. The summary table of Example 2.16 output states

```
source              1354342     3.7216E+00     8.1280E-01
```

There are 1354342 source tracks because 200000 source histories were requested (SSR run) from the initial 100000 SSW run: 2 × 677171 = 1354342. The average source weight is 3.7216E+00, and the average source energy is 8.1280E–01 MeV based on a source weight of 1.0 and 100000 histories in the initial SSW run. All other summary information and tallies in the output file are also based on the initial source weight of 1.0 and 100000 histories. The results of the current tally in the SSR run are

```
1tally          1          nps =       99271          Neutrons crossing
surfaces
number of histories used for normalizing tallies =       100000.00
          surface  a is (52.1 52.2 52.3)
          surface  b is (53.1 53.2 53.3)

   angle :       0.0000E+00          1.0000E+00          total surface
     a    1.29752E+00 0.0066  0.00000E+00 0.0000  1.29752E+00 0.0066
     b    0.00000E+00 0.0000  2.41864E+00 0.0062  2.41864E+00 0.0062
```

The SSR run results are statistically the same as SSW run results. For every 1 starting neutron, 1.297 neutrons cross surface 52 from the aluminum into the air, where they are killed. Then 3.7126 neutrons enter from the air on surface 52 in the SSW run and from the surface source in the SSR run. Then 2.419 cross surface 53 to escape the geometry.

In this example, the SSR run (Example 2.16) had the same result as the SSW run model of the entire geometry (Example 2.15), but the Figure of Merit (FOM) was 4.5 times higher—that is, the same result was achieved 4.5 times faster. The FOM is described in detail in Sect. 4.1 on Variance Reduction and is the sole measure of relative problem efficiency.

Following is an example of the SSW/SSR CEL option and how a KCODE criticality calculation can be on a simple cell that is then translated to multiple locations. A single uranium can has its KCODE criticality source neutrons written to a surface source file with the SSW CEL option in Example 2.17. Then this file is read in the four-can geometry with the SSR CEL option and translated to all four can locations in Example 2.18. The result ignores the small backscattered induced fission, but is a good approximation of running the KCODE calculation on the four-can geometry in Example 2.14.

Example 2.17 New Geometry Consisting of a Single Uranium Can from Example 2.14. The SSW CEL Option Records All Fission Neutrons from All Active Cycles of the Criticality Calculation in the One Uranium Can, Cell 24. Because the Can Is Surrounded by a Zero-Importance Cell, No Neutrons from Other Cans Enter. The Resulting Neutrons from Fission Sources Are Written to the SSW Surface Source Write File

```
Single uranium can in air and aluminum
24 101 -18.0    -50              imp:n=1  $ uranium can
23   0           50              imp:n=0  $ outside

50 RCC  0  0 -20  0 0 40   8     $ fuel cylinder

m101  92238 -.8  92235 -.2 nlib=.80c
print
kcode 1e4 1.0 30 130
sdef  axs=0 0 1 rad=d31 ext=d32
si31  0 8
sp31 -21 1
si32 -20 20
sp32   0 1
SSW   CEL 24
```

Example 2.18 Four-Can geometry of Example 2.14 with the Fission Source from One-Can Geometry Example 2.17, Read with SSR CEL and Translated to All Four Can Positions. The Approximation Introduced by the NONU Card Is Discussed Below

```
Four uranium cans in air and aluminum
11 101 -18.0      -51                     imp:n=1  $ uranium can
12 like 11 but trcl=(20  0 0)            imp:n=1  $ uranium can
13 like 11 but trcl=( 0 20 0)            imp:n=1  $ uranium can
14 like 11 but trcl=(20 20 0)            imp:n=1  $ uranium can
21 102  -1.0  -52 #11 #12 #13 #14        imp:n=1  $ air tank
22 103  -2.7  -53 52                     imp:n=1  $ Al liner
23  0              53                     imp:n=0  $ outside

51 RCC  0  0 -20  0 0 40    8                      $ fuel cylinder
52 RCC 10 10 -21  0 0 42   25                      $ air cylinder
53 RCC 10 10 -22  0 0 44   40                      $ aluminum liner

m101   92238 -.8   92235 -.2 nlib=.80c
m102   7014.80c .8   8016.80c .2
m103   13027.80c 1
print
nps    100000
NONU
SSR   CEL 24  TR=D40
SI40  L 0   42   43   44
SP40     1 3R
TR42   20  0 0
TR43    0 20 0
TR44   20 20 0
FC1:n Neutrons crossing surfaces
F1:n  (51.1 51.2 51.3) (52.1 52.2 52.3) (53.1 53.2 53.3)
C1      0 1 T
FQ1    F C
FC4   Fissions and neutron production in uranium cells
F4:n   11 12 13 14 T
FM4  (-1 101 -6) (-1 101 -6 -7)
FQ4    F M
SD4    1 4r
```

The surface tally output of Example 2.18 follows:

```
1tally        1       nps =      1296681   Neutrons crossing surfaces
number of histories used for normalizing tallies =      1000000.00
           surface  a is (51.1 51.2 51.3)
           surface  b is (52.1 52.2 52.3)
           surface  c is (53.1 53.2 53.3)
```

```
angle  :   0.0000E+00                    1.0000E+00                     total
surface
  a    2.22166E-01 0.0022  4.40912E-01 0.0013  6.63078E-01 0.0015
  b    2.88259E-01 0.0019  8.16748E-01 0.0008  1.10501E+00 0.0011
  c    0.00000E+00 0.0000  5.27283E-01 0.0006  5.27283E-01 0.0006
```

The surface tally output of Example 2.14 follows:

```
1tally         1          nps =      99770   Neutrons crossing surfaces
number of histories used for normalizing tallies =      1000591.00
           surface  a is (51.1 51.2 51.3)
           surface  b is (52.1 52.2 52.3)
           surface  c is (53.1 53.2 53.3)


angle  :   0.0000E+00                    1.0000E+00                     total
surface
  a    2.10443E-01 0.0076  4.21114E-01 0.0052  6.31556E-01 0.0057
  b    2.95602E-01 0.0068  8.18688E-01 0.0039  1.11429E+00 0.0045
  c    0.00000E+00 0.0000  5.21861E-01 0.0036  5.21861E-01 0.0036
```

Example 2.14 used a KCODE fission source. Example 2.18 used the SSR surface source instead with the CEL option and ran eight times faster and demonstrated its efficacy by achieving nearly the same result.

The results are comparable. The differences are because

- induced neutrons in each can from neutrons scattering in from other cans or reflected are ignored. Consequently the surface 51 tally, surface "a," is 5% lower;
- neither KCODE criticality calculation was fully converged, and the relative errors of the SSR calculation, Example 2.18, assume the surface source file from Example 2.17 was perfectly converged;
- the single can KCODE calculation Example 2.17 included neutrons from cycles that were not fully settled; and
- the four-can KCODE calculation Example 2.14 had different numbers of fissions in each can even though, by symmetry, they should have been the same.

Other points of interest are as follows:

- The NONU option was required in Example 2.18 because all fission neutrons in the single can were already accounted for in the surface source. However, this approach is approximate because induced fissions from in-scattering from other cans or back scatter from the aluminum liner are neglected. To avoid this incorrect albedo neglecting induced fissions, the SSR should use the sources on the surface rather than the CEL option on the SSR card and then allow fissions in the cells by omitting the NONU card.
- The SSR cell name is CEL = 24 even though there is no cell 24 in the SSR run of Example 2.18 because of SSW CEL 24 in Example 2.17.

- The source can be moved and duplicated anywhere with TR on the SSR card.
- Repeating the single-can source from Example 2.17 in the four can locations of Example 2.18 ensures that the fissions are the same in each can, which they should be from symmetry.
- Use of the CEL option on SSW/SSR is usually approximate. But in cases where there are many fissionable sources that are only loosely coupled to each other the SSW/SSR with the CEL option can be a very effective approximation. In safeguards science applications include barrels of nuclear waste, arrays of tanks of enriched or depleted uranium, and in some cases, reactor fuel pins.

The advantages of using SSW and SSR in safeguards runs are as follows:

- Calculations may be faster and more efficient when separated into several parts.
- Calculations that require analog sampling, such as multiplicity counters and pulse-height detectors, cannot use variance reduction but can be separated into parts that can take advantage of variance reduction.
- Parameter studies, particularly small perturbations with the MCNP perturbation (PERT) capability, can be applied to the subsequent SSR problem without having to run the entire SSW run initial problem.
- Calculations with a similar source, such as a bremsstrahlung electron accelerator, may be separated from subsequent SSR runs, modeling a variety of different targets.

Some disadvantages of using SSW and SSR in safeguard runs are

- the surface source write/read files may be very large;
- care must be taken for albedo effects—re-entering particles should not be duplicated or ignored;
- relative errors are not propagated; SSR assumes the SSW file has zero relative error and, thus, the SSR relative errors are underestimated;
- there may be streaming effects—the only directions sampled in the SSR run are those written in the SSW run, which may miss directions of interest; and
- point detectors and DXTRAN variance reduction require a user-input approximate surface direction (PSC option on the SSR card); this approximation is needed; particle tracks read from an SSR file have no recollection of the source or collision preceding the surface crossing.

2.3.6   Checking sources

The MCNP output file, OUTP, is the first way to check the source. All too often, what the user thinks was requested in the input file is not what the code thinks it was told! Print Tables 10, 110, 128, and 170 specifically describe the source. The problem summary and weight balance tables also provide important source information. The descriptions in Sect. 2.4.1 should be reviewed every time the source is modified.

Where possible, a tally should be constructed that confirms the source. The SCD and SCX special tally treatments are often useful. Tallies are described in Sect. 2.4.

Finally, it is possible to plot source distributions with the TMESH mesh tally and special options of the FMESH mesh tally, which are described in Sect. 2.6.

2.4 Output and Tallies

2.4.1 Output Files

MCNP output files, OUTP, are generated by every MCNP run. The basics of MCNP generated files are given in Sect. 2.1.1.

The output file is organized into Print Tables. If the PRINT card is in the input file or the option "PRINT" appears on the MCNP execution line, then the maximum print of the OUTP is made. Otherwise, only the minimum set tables are provided (default). The PRINT card has the form

```
PRINT   110
```

The above input card will print the minimum set of tables plus table 110.

```
PRINT   -85 -86 -128 -70 -98
```

The above input card (MCNP code manual section 3.3.7.2.1) will print the possible tables except print tables 85, 86, 128, and 70. Tables 85 and 86 are the electron and charged particle range tables, which are very lengthy and unchanged from calculation to calculation unless materials are changed. Print Table 128 is the "lattice universe map," which not only requires a lot of storage but slows the calculation by recording every source, entering, and collision in every cell and every repeated structures and lattice sub-cell. See the example in Sect. 2.1.7, "Filled Cells: Universes." Print Table 70 gives the surface coefficients of quadratic and macrobody surfaces and is of interest in only some circumstances. Print Table 98, giving basic constants, takes up little space but is the same for every run.

All of the OUTP output file should be reviewed the first time a problem is run, and then the parts that change should be reviewed every time the input file is changed or the files that the MCNP code is reading—such as the RSSA file from the surface source—are changed. Some of the more important Print Tables are described as follows.

Source Check: Optional Print Table 10 lists all SDEF parameters, showing the defaults when they have not been set in the input file. It also describes each source distribution and its probability, along with the KCODE criticality source parameters and what is read from the surface source read file, RSSA. See the example in Sect. 2.3.5, "Surface Source Read and Write (SSR, SSW)."

Variance-Reduction Summaries: Print Table 20 describes the MCNP interpretation of the input file weight windows when present. Print Table 62 describes forced collision and exponential transform variance-reduction input. The Problem Summary, particle activity by cell (Print Table 126), and particle weight balance by cell (Print Table 130) also contain variance-reduction activity summaries. Print Tables 180 and 190 summarize the weight window generator for variance reduction. Print Table 190 summarizes weight window generation from multigroup fluxes from alternative codes, if used. Generated cell-based weight windows, if any, are provided at the very end of the output file.

Geometry/Mass Check: Print Tables 40, 50, and 60 describe the cell volumes, masses, nuclide fractions, and importances. Masses should be checked against the physical masses being modeled. Track length tallies (F4, F6, F7) are normalized by volumes and masses, and these normalizations should be verified. In repeated structures or lattices, only the master cell volume and mass are calculated and a multiple of them must be used for the tally normalizations.

Materials Check: Print Table 100 is essential reading. About 10% of all user errors are due to not having the desired materials. Print Table 100 lists what the MCNP code is using, which can change from platform to platform and from time to time depending on which computing platform is hosting the MCNP calculation. Also, the intuitive specification of nuclides—such as 6012 for carbon neutron cross sections—causes the MCNP code to use decades-old data rather than the correct 6000 elemental specification for modern ^{12}C. A listing of available data is in LA-UR-20709 [4] for the MCNP6.2 release, but the MCNP code uses whatever is in the user's XSDIR data directory, and Print Table 100 should always be checked to know which data are actually used.

Source Check: Optional Print Table 110 lists the basic parameters of the first 50 histories: position (x,y,z), cell, surface, direction (u,v,w direction cosines), energy, weight, and time. If the source consists of multiple particle types, then the particle type is provided. If repeated structures and universes are used, then the source position, cell, and direction are provided at each level of the universe, along with lattice index, if appropriate. The energy, weight, and time columns are not shown in the following, which is from Example 2.13:

```
print table 110


 nps par   x              y              z      cell lattice[i j k] surface  u           v           w

   1   p  1.000E+01  3.000E+00   0.000E+00  15              0   2.033E-01  7.180E-01  6.657E-01
   2  sf -2.104E+00  2.156E+00  -1.874E+00  14                  7.962E-01 -4.029E-01 -4.513E-01
      sf -1.039E-01  2.156E+00   1.258E-01  21[-1 0 -1]         7.962E-01 -4.029E-01 -4.513E-01
      sf -1.039E-01  2.156E+00   1.258E-01   1              0   7.962E-01 -4.029E-01 -4.513E-01
   3   p  1.000E+01  3.000E+00   0.000E+00  15              0  -5.363E-01  6.280E-01 -5.638E-01
   4   n  1.334E+01  1.033E+01  -1.825E+00  11                  8.493E-01  4.573E-01  2.637E-01
       n  3.446E-01  1.033E+01  -9.270E-02  22[-4 1  0]         8.493E-01  4.573E-01  2.637E-01
       n  3.446E-01  1.033E+01  -9.270E-02   1              0   8.493E-01  4.573E-01  2.637E-0
```

Problem Summary: This essential summary of particle and energy creation and loss balance should be reviewed in every calculation. It is provided for each particle type and provides a wealth of important information. The following problem summary is from Example 2.13:

neutron creation	tracks	weight (per source particle)	energy (per source particle)
source	898096	1.0000E+00	1.8331E+00
nucl. interaction	0	0.	0.
particle decay	0	0.	0.
weight window	0	0.	0.
cell importance	0	0.	0.
weight cutoff	0	2.4484E-03	9.2821E-05
e or t importance	0	0.	0.
dxtran	0	0.	0.
forced collisions	0	0.	0.
exp. transform	0	0.	0.
upscattering	0	0.	1.7092E-08
photonuclear	0	0.	0.
(n,xn)	765	7.4328E-04	4.8758E-04
prompt fission	259293	1.8546E-01	3.7613E-01
delayed fission	1945	1.4743E-03	7.7146E-04
prompt photofis	0	0.	0.
tabular boundary	0	0.	0.
tabular sampling	0	0.	0.
total	1160099	1.1901E+00	2.2105E+00

number of neutrons banked	472911
neutron tracks per source particle	1.2917E+00
neutron collisions per source particle	1.0777E+01
total neutron collisions	9678987
net multiplication	1.1150E+00 0.0005

neutron loss	tracks	weight (per source particle)	energy (per source particle)
escape	1039960	1.0468E+00	1.2701E+00
energy cutoff	0	0.	0.
time cutoff	0	0.	0.
weight window	0	0.	0.
cell importance	0	0.	0.
weight cutoff	13869	2.4638E-03	8.1124E-05
e or t importance	0	0.	0.
dxtran	0	0.	0.
forced collisions	0	0.	0.
exp. transform	0	0.	0.
downscattering	0	0.	8.9672E-01
capture	0	6.5082E-02	1.0656E-02
loss to (n,xn)	382	3.7116E-04	3.1146E-03
loss to fission	105888	7.5412E-02	2.9833E-02
nucl. interaction	0	0.	0.
particle decay	0	0.	0.
tabular boundary	0	0.	0.
elastic scatter	0	0.	0.
total	1160099	1.1901E+00	2.2105E+00

average time of (shakes)		cutoffs	
escape	3.4851E+07	tco	1.0000E+33
capture	3.2604E+07	eco	0.0000E+00
capture or escape	3.4719E+07	wc1	-5.0000E-01
any termination	3.2597E+07	wc2	-2.5000E-01

photon creation	tracks	weight (per source particle)	energy (per source particle)
source	400222	4.4563E-01	6.7254E+01
nucl. interaction	0	0.	0.
particle decay	0	0.	0.
weight window	0	0.	0.
cell importance	0	0.	0.
weight cutoff	0	0.	0.
e or t importance	0	0.	0.
dxtran	0	0.	0.
forced collisions	0	0.	0.
exp. transform	0	0.	0.
from neutrons	484434	9.9354E-01	2.0178E+00
bremsstrahlung	8435674	9.6553E+00	1.5243E+01
p-annihilation	624882	7.1252E-01	8.1705E-01
photonuclear	0	0.	0.
electron x-rays	0	0.	0.
compton fluores	0	0.	0.
muon capt fluores	0	0.	0.
1st fluorescence	3241292	3.9805E+00	5.3427E-01
2nd fluorescence	591308	7.3783E-01	2.4329E-02
cerenkov	0	0.	0.
(gamma,xgamma)	0	0.	0.
tabular sampling	0	0.	0.
prompt photofis	0	0.	0.
total	13777812	1.6525E+01	8.5891E+01

photon loss	tracks	weight (per source particle)	energy (per source particle)
escape	1447030	1.7452E+00	5.061E+01
energy cutoff	0	0.	3.3191E-03
time cutoff	0	0.	0.
weight window	0	0.	0.
cell importance	0	0.	0.
weight cutoff	0	0.	0.
e or t importance	0	0.	0.
dxtran	0	0.	0.
forced collisions	0	0.	0.
exp. transform	0	0.	0.
compton scatter	0	0.	3.9870E+00
capture	12018341	1.4424E+01	3.3933E+00
pair production	312441	3.5626E-01	2.7897E+01
photonuclear abs	0	0.	0.
loss to photofis	0	0.	0.
total	13777812	1.6525E+01	8.5891E+01

```
number of photons banked              10136298        average time of (shakes)              cutoffs
photon tracks per source particle    1.5341E+01          escape            3.1190E+08          tco  1.0000E+33
photon collisions per source particle 1.6852E+01         capture           2.3323E+08          eco  1.0000E-03
total photon collisions                1513406           capture or escape 2.4173E+08          wc1 -5.0000E-01
                                                         any termination   2.4404E+08          wc2 -2.5000E-01

computer time so far in this run     1.29 minutes      maximum number ever in bank    94
computer time in mcrun               1.24 minutes      bank overflows to backup file   0
source particles per minute          8.0335E+05
random numbers generated              33745313         most random numbers used was         9187 in history    689906
```

Physics Summary: Optional Print Table 126 provides the activity in each cell for each particle type. The eight columns of information can also be plotted using the PAC, N, and PAR commands on the right-hand side of the interactive geometry plotter. The eight columns of information are:

PAC1: tracks entering (including source particles and secondary particles)
PAC2: population (does not include re-entrant particles)
PAC3: collisions
PAC4: collisions * weight
PAC5: number weighted average energy
PAC6: flux weighted average energy
PAC7: average track weight
PAC8: average track mean free path (MFP)

These quantities can be plotted after a run using the RUNTPE continuation file:

```
mcnp6 i=inp r=runtpefile z
```

or during the course of a run using the interrupt:

```
<ctrl-c><m>
```

Once the MCPLOT prompt appears, type "PLOT" to get the geometry plot. The command "LABEL 0 1 PAC3" in the bottom left block will put the number of collisions in each cell of the plot. Alternatively, click PAC, N, N, N, and L2. Click PAR to get the correct particle type if there is more than one.

Repeated Structures Physics Summary: Optional Print Table 128 provides the activity in each repeated structure/lattice element. A sample from Example 2.13 follows:

```
1neutron   activity in each repeated structure / lattice element
print table 128


        source     entering  collisions              path

        12371        23463        7105     1 < 21[-3 0 -3] <    14
            0        36450        2045     2 < 21[-3 0 -3] <    14
            0        53043       39364     3 < 21[-3 0 -3] <    14
        12338        29259        9835     1 < 21[-2 0 -3] <    14
            0        49460        2789     2 < 21[-2 0 -3] <    14
            0        75996       65336     3 < 21[-2 0 -3] <    14

        12467        29552       10042     1 < 21[-2 0  3] <    14
            0        50204        2786     2 < 21[-2 0  3] <    14
            0        77396       72336     3 < 21[-2 0  3] <    14
```

2677	6333	1304	1 < 22[1 4 0] <	11
0	7440	360	2 < 22[1 4 0] <	11
0	11781	6326	3 < 22[1 4 0] <	11
0	1125238	0	15	
0	0	0	16	

898096	18269879	9678987

Cell Weight Balance Summary: Optional Print Table 130 is the particle weight balance in each cell. The particle weight information from the summary table is provided for each individual cell in the problem.

Nuclide Activity by Cell Summary: Optional Print Table 140 is the activity of each nuclide in each cell, which includes the total collisions, collisions by weight, weight lost to capture, weight gain by fission, weight gain by (n,xn), and secondary particle production information. In coupled neutron-photon problems, it also includes several summary tables of the photoatomic activity of each nuclide in each cell, per source particle.

Source Check: Optional Print Table 170 describes each source distribution, how often it was sampled, and how often it was expected to be sampled. For surface source SSR runs, the RSSA file utilization is described.

The output file concludes with the tallies, which are described in the rest of Sect. 2.4. The tallies are user-specified and so general that they can be used to answer almost any question of what is going on in the calculation. For each tally, there is a tally fluctuation chart at the end of the output file and a "yes" or "no" answer to whether the tally passes each of ten statistical tests for convergence. There is also an optional more detailed statistical analysis of tally convergence (Print Table 160), optional printed plots of the probability density function (Print Table 161), and cumulative normed and unnormed tally density function plots (Print Table 162).

2.4.2 MCNP Estimators and Tally Types

The Monte Carlo method does not give answers. It provides estimates with statistical errors. In the MCNP code, the statistical errors are relative error fractions—one standard deviation of the mean divided by the mean value. The four possible estimators are as follows:

- Track length estimator: flux = weight * track length/volume. The MCNP code uses the track length estimator for cell-averaged flux, heating, k_{eff} criticality, and reaction rates.
- Surface estimator: the MCNP code uses the surface estimator for the current or flux of particles crossing a surface, energy balance, and pulse counts in cells.
- Collision estimator: the MCNP code uses the collision estimator for reaction rates, k_{eff} criticality, problem summary and energy deposition.

- Pseudo-deterministic or next-event estimators: the MCNP code uses next-event estimators for flux and reaction rates.

Flux vs. Fluence: Fluence is the time-integrated flux. The MCNP code is fully time-dependent even if time information is not requested in the tallies. If the source units are particles per second, then the tally quantity is flux in units of particles/cm^2/s. If the source is one source particle over all time, then the tally quantity is fluence, or particles/cm^2; that is, the tally units are determined by the source. If the source is in particles/s, then the tally will be per second.

Tally Types: The MCNP tally types are:

Tally	Type	Estimator	Fn Units	*Fn Units
F1:	Surface Current	W	#	MeV
F2:	Surface Fluence	$W/(|\mu|*A)$	#/cm^2	MeV/cm^2
F4:	Cell Fluence	$W\lambda/V$	#/cm^2	MeV/cm^2
F5:	Detector Fluence	$Wp(\mu)e^{-\alpha x}/2\pi R^2$	#/cm^2	MeV/cm^2
F6:	Energy Deposition	$W\lambda\sigma_T(E)H(E)\rho_a/\rho_g$	MeV/g	jerks/g
F7:	Fission Energy Deposition	$W\lambda\sigma_f(E)Q\rho_a/\rho_g$	MeV/g	jerks/g
F8:	Pulse Height	W_S in bin $E_D \bullet W/W_{Ss}$	pulses	MeV

The symbols are as follows:

W = particle weight; Ws = source weight
E = particle energy (MeV); E_D = Energy deposited (MeV)
μ = cosine of angle between surface normal and trajectory
A = surface area (cm^2); V = cell volume (cm^3)
λ = track length (cm)
$p(\mu)$ = probability density function: μ = cosine of angle between particle trajectory and detector
σ = total mean free path to the detector
R = distance to the detector (cm)
$\sigma_T(E)$ = microscopic total cross section (barns)
$H(E)$ = heating number (MeV/collision)
ρ_a = atom density (atoms/barn-cm)
ρ_g = gram density (g/cm^3)
M = cell mass (g) = $\rho_a * V$
$\sigma_f(E)$ = microscopic fission cross section (barns)
Q = fission heating Q-value (MeV)

The historical (Manhattan Project) unit "jerk" is equal to 10^9 Joules or 6.241×10^{21} MeV, or approximately one quarter of a metric ton of TNT.

- Track length estimator: F4, F6, F7, k_{eff}, reaction rates
- Surface estimator: F1, F2, F8, energy balance
- Collision estimator: F6, k_{eff}, reaction rates, energy balance
- Next-event estimator: F5 flux, reaction rates

2.4.3   Basic Tally Format

2.4.3.1   The Eight Tally Dimensions: FDUSMCET

There are eight dimensions of MCNP tallies:

- F: surface numbers (F1, F2), cell numbers (F4, F6, F7, F8), and point detector parameters (F5);
- D: surface flagging (SF), cell flagging (CF), or direct/total detector;
- U: user bins;
- S: segment bins;
- M: multiplier bins;
- C: angle bins;
- E: energy bins;
- T: time bins.

2.4.3.2   F: Surface and Cell Tallies

The basic tally format for cell and surface tallies (not F5 detector tallies) is

```
Fn:<pl>     S₁ S₂ . . . Sₖ
Fn:<pl>     S₁ (S₂ . . . S₃) (S₄ . . . S₅) S₆
```

where n is a positive integer with the following meaning. The last digit of n is the tally type. If it ends in 1, then it is an F1 tally. For example, F111:N. The particle type, $<pl>$, is N for neutrons, P for photons, E for electrons, H for protons, etc. The S_n entries are surfaces for F1 and F2 and cells for F4, F5, F6, F7, and F8. Parentheses can be used to sum over several surfaces or get the average over several cells. If the surfaces are macrobodies, the facets must be specified, as in Example 2.14, Example 2.15, and Example 2.18.

```
F1:n  (51.1 51.2 51.3)  (52.1 52.2 52.3)  (53.1 53.2 53.3)
F4:n   11 12 13 14 T
```

The T at the end of the F4 tally averages over all cells, which is equivalent to

```
F4:n   11 12 13 14  (11 12 13 14)
```

In some earlier versions of the MCNP code, surfaces—and particularly macrobody facets—that are eliminated for being the same as others cannot tallied upon.

2.4.3.3   F: Tallies in Lattices and Repeated Structures

Tallying in lattices and repeated structures requires specifying not only the cells or surfaces to be tallied but also the paths to them through various levels of universes. Example 2.19, which follows, uses the lattice/repeated structure geometry of Example 2.13. This geometry is illustrated in Fig. 2.9. Surfaces 51, 52, and 53 have been added so that both the surface tally 901 and energy deposition tally 906 can illustrate tally segmenting.

Example 2.19 Repeated Structures/Lattice Geometry from Example 2.12 and Example 2.13 Illustrating Tallies in Repeated Structures and Lattices and Segment Tallies. (Note: The MCNP6 Manual (Appendix C) Recommended Parameters for the ^{252}Cf Neutron Energy Watt Spectrum Are 1.18 (MeV) and 1.03419 (MeV^{-1})).

```
Square and hex lattices
1   313 -10.44    -1        u=1    imp:n=1  $ UO2
2   201  -6.55    -2 1      u=1    imp:n=1  $ Zr clad
3   609  -1.00       2      u=1    imp:n=1  $ H2O
21 111  -4.00    -31 lat=1 u=11  imp:n=1  $ Square lattice
             fill = -3:3 0:0 -3:3
                          1 1 1 1 1 1 1
                          1 1 1 1 1 1 1
                          1 1 1 1 1 1 1
                          1 1 1 1 1 1 1
                          1 1 1 1 1 1 1
                          1 1 1 1 1 1 1
                         11 1 1 1 1 1 1
22 111  -4.00    -32 lat=2 u=12  imp:n=1  $ Hexagonal lattice
             fill =   -5:5 -4:4 0:0
                        0 0 0 0 1 1 1 1 1 1 1
                        0 0 0 1 1 1 1 1 1 1 1
                        0 0 0 1 1 1 1 1 1 1 0
                        0 0 1 1 1 1 1 1 1 1 0
                        0 0 1 1 1 1 1 1 1 0 0
                        0 1 1 1 1 1 1 1 1 0 0
                        0 1 1 1 1 1 1 1 0 0 0
                        1 1 1 1 1 1 1 1 0 0 0
                       12 1 1 1 1 1 1 0 0 0 0
14   0    -4                fill 11 imp:n=1  $ Square lattice box
11 like 14 but TRCL=(20 0 0) fill 12        $ Hexagonal lattice box
15   0    #14 #11 -20                imp:n=1  $ outside boxes
16   0            20                 imp:n=0  $ outside world
```

```
1    rcc  0 0 0   0 12 0   .5      $ Right circular cylinder
2    rcc  0 0 0   0 12 0   .6      $ Right circular cylinder
4    rpp -7 7    0 12  -7 7        $ Right parallelepiped 14x12x14
31   rpp -1 1    0 12  -1 1        $ Right parallelepiped 2x12x2
32   rhp  0 0 0  0 12 0  1 0 0     $ Right hex 12x1
20   sph 0 0 0 50                  $ Sphere containing lattice boxes
51 PX  0
52 PX  8
53 PX 12


MODE N P
m0 nlib=.70c plib=.04c elib=.03e
m201   40090 -0.505239   40091 -0.110180   40092 -0.168413
       40094 -0.170672   40096 -0.027496   nlib=80c
m313   92235 -0.02759   92238 -0.85391   8016 -0.11850
M609   1001 2   8016 1
MT609  LWTR.10t
M111   6000 1   1001 1   7014 1
MT111  POLY.10t BENZ.10t GRPH.10t
SDEF   PAR=D42
       CEL=FPAR=D51  POS=FPAR=D53  RAD=FPAR=D54  EXT=FPAR=D55
       AXS=FPAR=D56  TME=FPAR=D57  ERG=FPAR=D58
SI42   L  SF   N    P    P    P    P                         $ PAR
SP42      .3   .3   .1   .1   .1   .1                        $ PAR
DS51   S  46   47   48   48   48   48                        $ CEL
DS53   L  0 6 0   0 6 0  10 3 0  10 3 0  10 7 0  10 7 0      $ POS
DS54   S  43   43   0    0    0    0                         $ RAD
DS55   S  44   44   0    0    0    0                         $ EXT
DS56   S  45   45   0    0    0    0                         $ AXS
DS57   L  0 1e8 2e8 3e8 4e8 5e8                              $ TME
DS58   S  0   92   93   94   95   96                         $ ERG
SC46   Spontaneous fission cells in XYZ lattice
SI46   L (1<21[-2:3 0 3]<14) (1<21[-3:3 0:0 -3:2]<14)
SP46      1 5R                  1 41R
SC47   Neutron source cells in HEX lattice
SI47   L (1<22[-5    3  0]<11)
          (1<22[-4  1:4  0]<11) (1<22[ -3 -1:4  0]<11)
          (1<22[-2 -3:4  0]<11) (1<22[-1:1 -4:4  0]<11)
          (1<22[ 2 -4:3  0]<11) (1<22[   3 -4:1  0]<11)
          (1<22[ 4 -4:-1 0]<11) (1<22[   5 -4:-3 0]<11)
SP47      .5
          .5 1 1 .5                    .5 1 3R .5
          .5 1 5R .5 .5 1 6R .5 .5 1 6R .5 .5 1 6R .5
          .5 1 5R .5                   .5 1 3R .5
          .5 1 1 .5                    .5 .5
```

```
SC48    Point photon source cell in water between lattices
SI48    L  15
SP48        1
c       particle-dependent radius, extent, axis
SI43    0 .5                                              $ Radii
SP43   -21  1                                             $ Radii
SI44    -6  6                                             $ Extent
SP44   -21  0                                             $ Extent
SI45    L   0 1 0                                         $ Axis
SP45        1                                             $ Axis
c       energies dependent upon time, position, and particle type
SP92    -3 1.175 1.0401              $ 252Cf Frohner Watt parameters
SI93    50 59 $ Photon, XYZ water, TME 2e8, energies
SP93     0  1
SI94    60 69 $ Photon, XYZ water, TME 3e8, energies
SP94     0  1
SI95     L 70 $ Photon, XYZ water, TME 4e8, energies
SP95        1
SI96     L 80 $ Photon, XYZ water, TME 5e8, energies
SP96        1
nps  100000
PRINT -162 -85 -86 -140
c ***** Tallies **************
FC901 Segmented surface tally
F901:n (4.1 4.2) (11004.1 11004.2) (2.1<21<14) (2.1<22<11)
        (11004.1 4I 11004.6) (2.1<21[0 0 0]<14)
FS901 -52 -53 T               $ 4 segment bins
FQ901 F S                     $ Print cells down, segments across
FC906 Segmented energy deposition tally
+F906 (1<21<14) (1<22<11)         (21<14) (22<11)
        (1<21[-3:3 0:0 -3:3]<14)  (1<22[-5:5 -4:4 0:0]<11)
FS906 -51 -52 -53 T           $ 4 segment bins
FQ906 F S                     $ Print cells down, segments across
SD906 1 759R                  $ MeV units, 102 cells x 5 segments
```

Tallying in lattices and repeated structures is illustrated in Example 2.19 with the surface tally F901 and cell tally F906.

The "<" separates different universe levels. The cells or surfaces on each side must be in the universe at the next level and not in the same universe.

The F901 tally provides the format for F1 and F2 type surface tallies in lattices or repeated structures. In this example, it counts neutrons in the following bins:

- Bin 1 (4.1 4.2): either facet 1 or 2 of surface 4;
- Bin 2 (11004.1 11004.2): either facet 1 or 2 of surface 4 translated to cell 11;

- Bin 3 (2.1<21<14): any of the 49 appearances of surface 2.1 u=1 filling rectangular lattice cell 21 u=11, which fills cell 14 u=0;
- Bin 4 (2.1<22<11): any of the 99 appearances of surface 2.1 u=1 filling hexagonal lattice cell 22 u=12, which fills cell 11 u=0;
- Bin 5 (11004.1 4I 11004.6): surfaces 4.1 and 4.2 of surface 4 translated to cell 11. In addition, bin 5 also counts all neutrons crossing surfaces 1.2, 1.3, 4.5, and 4.6. because these surfaces are the same as 11004.3, 11004.4, 11004.5, and 11004.6. Further, the particles that cross surfaces 1.2, 1.3, 4.5, and 4.6 include particles that cross these surfaces anywhere in the problem and not just in cell 11; and
- Bin 6 (2.1<21[0 0 0]<14): surface facet 2.1 u=1 filling lattice element 0,0,0 in cell 21 u=11 filling cell 14 u=0.

The +F906 tally has no particle type because +F6 energy deposition tallies include the energy deposition from all particle types in the problem from whichever estimator they use—either track length or collision. The F906 tally provides the format for F4, F6, and F7 type cell tallies in lattices or repeated structures. Again, the "<" separates universe levels; in this example, it tallies the energy deposition in the following bins from:

- Bin 1 (1<21<14): the sum of all 49 occurrences of cell 1 u=1 filling cell 21 u=11 filling cell 14 u=0;
- Bin 2 (1<22<11): the sum of all 99 occurrences of cell 1 u=1 filling cell 22 u=12 filling cell 11 u=0;
- Bin 3 (21<14): the 1 occurrence of cell 21 u=11 filling cell 14 u=0;
- Bin 4 (22<11): the 1 occurrence of cell 22 u=12 filling cell 14 u=0;
- Bins 5–53 (1<21[-3:3 0:0 -3:3]<14): each of the 49 occurrences of cell 1 u=1 filling cell 21 u=11 filling cell 14 u=0; and
- Bins 54–152 (1<22[-5:5 -4:4 0:0]<11): each of the 99 occurrences of cell 1 u=1 filling cell 22 u=12 filling cell 11 u=0.

Note that bins 3 and 4, (21<14) and (22<11), are needed because the (1<21[-3 0 3]<14) bin in which cell 21 is filled with itself and the (1<22[-5 4 0]<11) bin in which cell 22 is filled with itself have zero tallies because cell 1 is not in these lattice positions.

The U=n notation (not shown) may also be used at any level when there are multiple cells at that level, but this capability does not work in some cases, such as the above (see the MCNP code manual section 3.3.5.1.4).

2.4.3.4   C: Angle Bins and FC Tally Comments

Surface tallies can have angle bins and nearly all tallies can have time and/or energy bins. Also, a descriptive tally title is strongly encouraged. From Example 2.14, Example 2.15, and Example 2.18:

```
FC1:n Neutrons crossing surfaces
F1:n (51.1 51.2 51.3) (52.1 52.2 52.3) (53.1 53.2 53.3)
C1    0 1 T
```

The title on the tally comment card, FC1, has the description "Neutrons crossing surfaces" everywhere in the output or in a plot where tally 1 occurs.

The angle of particles that cross the surface is specified by the C1 cosine card. The cosine bins are $-1 < \mu < 0$, $0 < \mu < 1$, and $-1 < \mu < 1$ from the "T" option, which is the total over all other bins. The cosines are relative to the surface outward normal. Thus, the cosine bins are everything crossing inward, everything crossing outward, and everything crossing in either direction. If *C1 is used, then the angles are in degrees, and the specification goes from 180°. For example, one-degree angle bins would be

```
*C1    179  177I  0   T
```

The lower limit of the bins, 180°, is not specified. Between 179° and 0°, there are 177 interpolated points.

2.4.3.5 E, T: Energy and Time Bins, Default Bins, and FQ and TF Format Control

Energy bins are specified in MeV with the upper energy bound, as in Example 2.20 for the heating tally F16:

```
e16   0.2 48I 10
```

The total over all energy or time—the "T" option—is the default for time and energy bins and can be turned off with the "NT" option. The above is the same as

```
e16   0.2 48I 10 T
```

Example 4.1 shows a number of capabilities:

```
t0   1e2 263ilog 2e5
e0   1e-4 1e-3 1e-2 1e2
fq0 T E
```

The time bins are in shakes (10^{-8} s) and go from 1 μs to 2 ms with logarithmic interpolation. The default energy bins are one bin over all energy; the default time bins are one bin over all time. The E0 and T0 change the default for all tally time and energy bins unless they are set for specific tallies, as in e16 above.

By default, energy bins are printed vertically in the output file and time bins horizontally. The FQ0 changes the default for all tallies to print the time bins down and the energy bins across. These are then in blocks of C bins, in blocks of M bins, in blocks of S bins, in blocks of U bins, in blocks of D bins, and in blocks of F bins. The FQ changes the order of printing (see MCNP manual section 3.3.5.6).

From Example 2.14, Example 2.15, and Example 2.18:

```
FQ1     F C
FQ4     F M
```

Tally card FQ1 changes the default from

```
        F   D   U   S   M   C   E   T
```

to

```
FQ1     D   U   S   M   E   T   F   C
```

Here, the output file prints C bins across and F bins down within blocks of T bins within blocks of E bins, etc.

Tally card FQ4 changes the default to

```
FQ1     D   U   S   C   E   T   F   M
```

Here, the output file prints M bins across and F bins down within blocks of T bins within blocks of E bins, etc.

The fq0 T E input of Example 4.1 (in Sect. 4.1.2) prints energy bins across and time bins down.

The eight tally dimensions are also used with the tally fluctuation chart control, TF, which points to which bin of a tally is to be used for the tally fluctuation chart at the end of the output. The TF card bin is used to obtain the statistical analysis of a quantity of real interest in the simulation (for example, the user may not be interested in the last time bin of a problem) but also is crucial for the weight window generation (Sect. 4.1) target tally bin. The default TF entries are

```
TFn     1   1   L   L   1   L   L   L
```

These entries point to the 1st F (cell, surface, or detector) bin, 1st D (flagging) bin, last U (user) bin, last S (segment) bin, 1st M (multiplier) bin, last C (cosine) bin, last E (energy) bin, and last T (time) bin.

In Example 3.4:

```
t4      10e8  98i  1000e8
tf4     7j  7
```

The above TF4 card uses the default F, D, U, S, M, C, E bins but the 7th T bin instead of the last 101st total bin.

2.4.3.6　M: Multiplier Bins and SD Divisors

The multiplier bins, which are the M dimension of tallies, are often used to convert flux tallies to reaction rates, were described in Example 2.14:

```
FC4   Fissions and neutron production in uranium cells
F4:n   11 12 13 14 T
FM4  (-1 101 -6)  (-1 101 -6 -7)
FQ4    F M
SD4    1 4r
```

Tally 4 in Example 2.14 multiplies the flux in each fissionable can by $\rho\sigma_f$ and $\rho\nu\sigma_f$ to get the number of fissions and number of fission neutrons produced. The basic format of the tally multiplier (FM) card is

```
FMn  (C₁ m₁ R₁)  (C₂ m₂ R₂) . . . T
```

n = tally number

C_i = multiplicative constant (if $C_i = -1$, for type 4 tallies only, the tally is multiplied by the atom density of the cell, ρ_a)

m_i = material number identified on an Mm card

R_i = a combination of ENDF reaction numbers

The reaction number values follow the ENDF (evaluated nuclear data file) standard and have the special MCNP values, as shown in Table 2.1:

Table 2.1 The Principal MCNP Reaction Number (MT) Values (PN is photonuclear; rxn is reaction). The S(α,β) reactions are: 1 = total, 2 = elastic, 4 = inelastic. The neutron cross sections do not include thermal effects. The photon cross sections do not include form factors for coherent and incoherent scattering

Neutrons		Photons		Protons	
1	Total	−1	Incoherent	1	Total
−2	Absorption	−2	Coherent	2	Non-elastic
−4	Heating	−3	Photoelectric	3	Elastic
−5	Gamma prod'n	−4	Pair production	4	Heating
−6	Total fission	−5	Total	>4	Other rxns.
−7	fission ν	−6	Heating	i00R	Particle i from rxn R
−8	fission Q	1	PN total		
16	(n,2n)	2	PN non-elastic		
17	(n,3n)	3	PN elastic		
18	(n,f)	4	PN heating		
103	(n,p)	>4	PN other rxns.		
		i00R	PN particle i from rxn R		

Other examples are

```
F2:N       1    2 $ 36 tally bins
FM2        (1.0) (2.0) (3.0) $ Constant multipliers
E2         .5 1 2 4 10 T $ Energy bins
F4:N       (1 2) 3 T$ 6 tally bins
FM4        (-1 1 -6 -7) $ Track-length estimate of k_eff
           (-1 2  1 -4)$ Neutron Heating (MeV/cm3)
```

The SD4 1 4r in the above from Example 2.14 changes the units by overriding the normal tally divisor. The SD card replaces the usual divisor for F (cell/surface) and S (segment) bins. The usual divisors for F2, F4, F6, and F7 are Area, Volume, Mass, and Mass. Thus, "SDn 1" would divide a tally bin by unity instead of the area (for an F2 tally) effectively multiplying the tally by the area. The new units would be particles, particles, MeV, and MeV rather than $1/cm^2$, $1/cm^2$, MeV/g, and MeV/g. The number of SD entries is the number of cells or surfaces on the F card times the number of segment bins.

2.4.3.7   D: Direct and SF, CF Flagging Bins

The D bins are described along with point detector tallies in Sect. 2.4.6. They are also used for surface, SF, and cell, CF, flagging. For example,

```
SF4   11.3   12
```

Tally 4 then has two D bins—the total, or usual tally, and that part of the tally that was made after particle tracks crossed surfaces 11.3 and 12.
For example,

```
CF1   20   21
```

Tally 1 then has two D bins—the total, or usual tally, and that part of the tally that was made after particle tracks entered cells 20 or 21.

2.4.3.8   U: User Bins

User bins, FU, enable input into tallies programmed by users who are altering the MCNP source code. They are more commonly used in conjunction with special tally treatments described in Sect. 2.4.4.

2.4.3.9   S: Segment Bins

Tally segmenting is shown in Example 2.19. The tally portion of that example is repeated here:

```
c ***** Tallies **************
FC901 Segmented surface tally
F901:n (4.1 4.2) (11004.1 11004.2) (2.1<21<14) (2.1<22<11)
      (11004.1 4I 11004.6) (2.1<21[0 0 0]<14)
FS901 -52 -53 T            $ 4 segment bins
FQ901 F S                  $ Print cells down, segments across
FC906 Segmented energy deposition tally
+F906 (1<21<14) (1<22<11)           (21<14) (22<11)
      (1<21[-3:3 0:0 -3:3]<14)   (1<22[-5:5 -4:4 0:0]<11)
FS906 -51 -52 -53 T        $ 5 segment bins
FQ906 F S                  $ Print cells down, segments across
SD906 1 759R               $ MeV units, 152 cells x 5 segments
```

Note that surfaces 51, 52, and 53

```
51 PX   0
52 PX   8
53 PX  12
```

are used only for segmenting; they are not part of the problem geometry because these surfaces do not appear within the cell card section of the input. They only appear on FS tally segmenting cards.

```
FS901 -52 -53 T            $ 4 segment bins
```

FS901 causes the surfaces of tally 901 to tally in 4 segment bins:

Segment bin 1: −52: all surface crossings with $x < 8$ cm, namely the square lattice;
Segment bin 2: −53: all surfaces not in the first bin but crossing with $8 < x < 12$, namely the geometry between the square lattice and the hexagonal lattice;
Segment bin 3: everything else bin, which would include the hexagon lattice;
Segment bin 4: total over all segments, "whole," which is the same tally that F901 would give without FS901 segment bins.

The F901 tally output is

```
1tally        901        nps =       100000
    +                                   Segmented surface tally
              tally type 1    number of particles crossing a surface.
              particle(s): neutrons
              surface  b is (11004.1 11004.2)
              surface  e is (11004.1 11004.2 1.2 1.3 4.5 4.6)
              surface  f is (2.1<21[0 0 0]<14)
              segment  1:            -52
              segment  2:            52       -53
              segment  3:            52       53
              segment  4:    whole surface


       segment:           1                    2                    3                 whole
       surface
     (4.1 4.2) 2.55726E-01 0.0080  0.00000E+00 0.0000  0.00000E+00 0.0000  2.55726E-01 0.0080
         b      0.00000E+00 0.0000  0.00000E+00 0.0000  1.79995E-01 0.0087  1.79995E-01 0.0087
     (2.1<21<14) 3.62434E+00 0.0074  0.00000E+00 0.0000  0.00000E+00 0.0000  3.62434E+00 0.0074
     (2.1<22<11) 0.00000E+00 0.0000  0.00000E+00 0.0000  2.09843E+00 0.0082  2.09843E+00 0.0082
         e      5.02658E-01 0.0061  0.00000E+00 0.0000  4.58745E-01 0.0052  9.61403E-01 0.0038
         f      1.08731E-01 0.0184  0.00000E+00 0.0000  0.00000E+00 0.0000  1.08731E-01 0.0184
```

The 1st, 3rd, and 5th surface bins of F901 are in the square lattice and thus score in segment bin 1. The 2nd, 4th, and 6th surface bins of F901 are in the hexagonal lattice and thus score in segment bin 3. The surface tallies FS segment bins are applied at the u=0 "real world" universe level coordinate system, not the lower universe level coordinate system of the surfaces being tallied.

```
FS906 -51 -52 -53 T     $ 5 segment bins
```

FS906 causes the cells of tally 906 to tally energy deposition in five segment bins:

Segment bin 1: −51: all cell energy deposition with $x < 0$ cm, namely the $-x$ portion of the square lattice;

Segment bin 2: −52: all cell energy deposition not in the first bin but with $0 < x < 8$, namely the $+x$ portion of the square lattice;

Segment bin 3: −53: all cell energy deposition not in the first two bins but crossing with $8 < x < 12$, namely the geometry between the square lattice and the hexagonal lattice;

Segment bin 4: everything else bin, which would include the hexagon lattice;

Segment bin 5: total over all segments, "whole," which is the same tally that F906 would give without FS906 segment bins.

The F906 tally output has 760 bins, so only parts are shown here:

```
1tally    906    nps =    100000
 +                   Segmented energy deposition tally
           tally type 6+    energy deposition    units    mev/gram
           particle(s): neutrons
           cell e is (1<21[-3 0 -3]<14)
           cell f is (1<21[-2 0 -3]<14)
           cell g is (1<21[-1 0 -3]<14)

           cell eo is (1<22[-2 4 0]<11)
           cell ep is (1<22[-1 4 0]<11)
           cell eq is (1<22[0 4 0]<11)
           cell er is (1<22[1 4 0]<11)

                   :

                   :

 segment:          1                    2                    3                    4                    whole
 cell
 (1<21<14)  6.01161E+00 0.0113  6.85838E+00 0.0102  0.00000E+00 0.0000  0.00000E+00 0.0000  1.28700E+01 0.0101
 (1<22<11)  5.39556E+00 0.0105  3.81918E+00 0.0131  0.00000E+00 0.0000  0.00000E+00 0.0000  9.21474E+00 0.0107
 (21<14)    6.70886E+00 0.0106  7.93863E+00 0.0093  0.00000E+00 0.0000  0.00000E+00 0.0000  1.46475E+01 0.0093
 (22<11)    6.28464E+00 0.0099  4.47032E+00 0.0119  0.00000E+00 0.0000  0.00000E+00 0.0000  1.07550E+01 0.0100
 e          2.54071E-02 0.1040  3.18441E-02 0.0944  0.00000E+00 0.0000  0.00000E+00 0.0000  5.72512E-02 0.0869
 f          5.09948E-02 0.0925  5.45180E-02 0.0800  0.00000E+00 0.0000  0.00000E+00 0.0000  1.05513E-01 0.0753
 g          6.50667E-02 0.0749  6.09488E-02 0.0761  0.00000E+00 0.0000  0.00000E+00 0.0000  1.26016E-01 0.0678

 eo         1.56021E-02 0.1168  1.42059E-02 0.1184  0.00000E+00 0.0000  0.00000E+00 0.0000  2.98080E-02 0.1034
 ep         1.08991E-02 0.1180  1.17894E-02 0.1248  0.00000E+00 0.0000  0.00000E+00 0.0000  2.26685E-02 0.1062
 eq         9.31922E-03 0.1227  9.11473E-03 0.1529  0.00000E+00 0.0000  0.00000E+00 0.0000  1.84340E-02 0.1178
 er         4.57305E-03 0.1316  5.53235E-03 0.1952  0.00000E+00 0.0000  0.00000E+00 0.0000  1.01054E-02 0.1476

                   :
```

All energy deposition is scored in the 1st and 2nd segment bins, namely $x < 0$ and $0 < x < 8$. The hexagonal lattice cells that fill cell 11 with $x > 12$ do not score in the 4th segment bin with $x > 12$ because cell segmenting is performed at the universe level of the cell. Both the square lattice that fills cell 14 and the hexagonal lattice that fills cell 11 have their centers at the origin in the $u = 1$ lowest level universe. Therefore, cell segments apply to the coordinate system of the level of the cell being tallied.

The segments are printed horizontally and the surfaces and cells vertically because of the FQ901 and FQ906 tally format specifications.

SD906 has 760 entries: 152 cells × 5 segments. The value of 1 replaces the normal FD6 divisor of grams, changing the units from MeV/g to MeV.

2.4.4   Special Tally Treatments

Special tally treatments may be applied to various tallies:

```
Form:   FTn   id₁ p₁,₁ p₁,₂ … id₂ p₂,₁ p₂,₂… …
              n = tally number
              Id = Special tally treatments given below
              pᵢ,ⱼ = parameter j for the iᵗʰ tally treatment.
```

The special tally treatments are

- FRV—Fixed arbitrary reference direction for tally 1 cosine binning
- GEB—Gaussian energy broadening
- TMC—Time convolution
- INC—Identify the number of collisions
- ICD—Identify the cell from which each detector score is made
- SCX—Identify the sampled index of a specified source distribution
- SCD—Identify which of the specified source distributions was used
- ELC—Electron current tally
- PTT—Put different multigroup particle types in different user bins
- RES—Residual nuclei
- TAG—Tally tagging
- LET—Tally stopping powers instead of energy
- ROC—Receiver-operator characterization
- PDS—Point detector sampling
- FFT—First fission tally
- COM—Compton image tally
- CAP—Coincidence capture
- PHL—Pulse-height light tally

2.4.4.1 FRV: Fixed Arbitrary Reference Direction for Tally F1 or F2 Cosine Binning

```
FTn    FRV    u    v    w
```

The default cosine bins are relative to the surface normal. The FRV treatment makes the cosine binning relative to the vector (u,v,w).

2.4.4.2 GEB: Gaussian Energy Broadening

The tallying of a gamma-ray pulse height spectrum in the MCNP code, by default, is equivalent to using a detector with infinite pulse height resolution. This is a very useful feature for studying the properties of the incoming gamma-rays. For simulation of experiments it may be necessary to include the actual resolution of the detector, for example, in triggering coincidences (see Sect. 2.4.4.18). Also, a user may simply want the simulated spectrum to resemble an experimentally measured spectrum. In these cases the MCNP tally can be broadened with the built-in tally treatment, GEB, to simulate the finite resolution of a real detector (an example is given in Sect. 3.3.4). The variation of resolution of the detector as a function of energy is accommodated by three parameters. The usage is as follows:

```
FTn    GEB    a    b    c
```

The parameters specify the full width at half maximum (FWHM) of the observed energy broadening in a physical radiation detector: $FWHM(E) = a + b\sqrt{E + cE^2}$, where E is the energy of the particle. The units of a, b, and c are MeV, MeV$^{1/2}$, and 1/MeV, respectively. The energy actually scored is sampled from a Gaussian distribution with that FWHM (from MCNP manual Sect. 3.3.5.18).

2.4.4.3 TMC: Time Convolution

```
FTn    TMC    a    b
```

The tally scores are made as if the source were a square pulse starting at time a < time < b. All particles should be started at time zero in the source definition (SDEF).

2.4.4.4 INC: Identify the Number of Collisions

```
FTn    INC
FUn    0    1    8I 10 1e8 T
UNC:N    k
```

The FUn bins are the number of collisions that have occurred in the track. The first bin has uncollided particles; the second has those with exactly one collision; the third entry interpolates so that those particles having exactly 2, 3, 4, 5, 6, 7, 8, 9, 10 collisions are in those bins. The last entry, 1e8, should be a large number to count all collisions. The UNC card controls whether the collision count starts over for each secondary particle ($k = 1$, default) or continues ($k = 0$).

2.4.4.5   ICD: Identify the Cell from Which Each F5 Detector Score Is Made

```
FTn    ICD
FUn    11 12 13 … 88 89 T
```

The FUn bins are the names of some or all of the cells in the problem. Thus, the detector tally is subdivided into bins according to which cell had the source or collision that resulted in the detector score.

2.4.4.6   SCX: Identify the Sampled Index of a Specified Source Distribution

```
FTn   SCX   k
```

One user bin is created for each bin of source distribution SIk plus a total bin, thus subdividing the tally into parts corresponding to each bin of the source distribution.

2.4.4.7   SCD: Identify Which of the Specified Source Distributions Was Used

```
FTn    SCD
FUn    k₁   k₂   k₃    . . .
```

Each FUn user bin of the tally is the portion of the tally from source distribution k_i. If more than one of the source distributions listed on the FU card is used for a given source particle, only the first one used will score.

2.4.4.8   ELC: Electron Current Tally

```
FTn    ELC   c
```

ELC specifies how the charge of a particle is scored to an F1 tally. Without the ELC treatment, the F1 tally gives particle current without regard for the particle charge. The possible values for c are

- $c = 1$ to cause negatively charged particles to make negative scores,
- $c = 2$ to put charged particles and antiparticles into separate user bins, and
- $c = 3$ for the effect of both $c = 1$ and $c = 2$.

If $c = 2$ or 3, three user bins (e.g., positrons, electrons, and total) are created.

2.4.4.9 PTT: Put Different Multigroup Particle Types in Different User Bins

```
FTn    PTT
FUn    a₁   a₂   a₃    . . .
```

Each FUn user bin is the atomic weights in units of MeV of particles that masquerade as neutrons in a multigroup data library (MGOPT multigroup option turned on). Thus $a_1, a_2 = 0.511, 0$ separates the tally into electrons and photons.

2.4.4.10 RES: Residual Nuclei

```
Fn:#  a₁   a₂   a₃    . . .
FTn   RES  8016  20040  26000  92238
```

For heavy ion (PAR = #) surface current (F1), surface flux (F2), cell flux (F4), and cell heating (F6) tallies, $n = 1, 2, 4, 6$, the tally is subdivided into user bins according to which nuclide heavy ion, ^{13}O, ^{40}Ca, all Fe isotopes, and ^{238}U made the tally in each cell or surface a_1, a_2, a_3.

```
F8:#  a₁   a₂   a₃
FT8   RES  z₁   z₂
```

For F8 tallies, the weight of all ions with Z numbers $z_1 < z < z_2$ produced by any reaction with any particle in the problem are recorded. If z_1, z_2 are ZAID identifiers, then the production of those nuclides is tallied.

```
MODE   N  H
F128:# 1101   1102
FT128  RES   5  8
FQ128  U   F
```

This F128 tally type 8 would list the production of all possible isotopes of B, C, N, and O. The production would come from primary reactions with either neutrons (N) or protons (H) and not secondary reactions if the produced nuclide decays (requiring the ACT activation card). The FQ128 tally format card lists the cells (F) horizontally and residual nuclei (U) vertically.

```
F258:#    1103   9I   1113
FT258   RES   1002  1003   2003   2004
FQ258   F    U
```

This F258 tally type 8 lists the production of deuterons, tritons, helions, and alphas from all problem particle types in cells 1103–1113. The FQ258 format lists the four light ions horizontally and the eleven cells vertically in the output.

2.4.4.11   TAG: Tally Tagging

```
FTn    TAG   a
FUn    b₁   b₂   b₃   b₄   . . .
```

Tally tagging separates a tally into bins by how and where the scoring particle was produced and works for all but type 8 tallies. The TAG options are

a=1: collided particles lose their tag; bremsstrahlung and annihilation photons are included in the scattered FU 0 bin;

a=2: collided particles lose their tag; bremsstrahlung and annihilation photons are given special tags for segregation;

a=3: all collided particles retain their production tag (most useful option); and

a=4: all collided particles except Compton photoatomic retain their production tag.

Each FUn bin, b_i, has the form CCCCCZZAAA.RRRRR

CCCCC = cell number or 00000 to tag all cells

ZZAAA = target nuclide identifier

RRRRR = reaction identifier (e.g., 00102 for n,γ) or residual nuclide ZAID

Tally tagging FUn bin, b_i, special cases:

−0000000001 or −1 source particle tag for all cells

−CCCCC00001 source (i.e., uncollided) particle tag for cell CCCCC

0000000000 or 0 scattered particle tag

10000000000 or 1e10 everything else tag

Tally tagging FUn bin, b_i, special cases for neutrons:

ZZAAA.99999 delayed particles from fission or residuals of ZZAAA

Tally tagging FUn bin, b_i, special cases for photons:

00000.00001	bremsstrahlung from electrons
ZZ000.00003	fluorescence from nuclide ZZ
00000.00003	K X-rays from electrons
00000.00004	annihilation photons from e−
ZZ000.00005	Compton photons from nuclide ZZ
ZZAAA.00006	muonic X-rays from nuclide ZZAAA
00000.00007	Cerenkov photons

Tally tagging FUn bin, b_i, special cases for electrons:
Electron special designations for ZZAAA.RRRRR:

ZZ000.00001	photoelectric from nuclide ZZ
ZZ000.00003	Compton recoil from nuclide ZZ
ZZ000.00004	pair production from nuclide ZZ
ZZ000.00005	Auger electron from nuclide ZZ
00000.00005	Auger electron from electrons
00000.00006	knock-on electrons

Tally tagging photon example:

```
FT5    TAG  3
FU5    -1.0                 $ Source from photons
       0000106012.00005     $ Compton from carbon in cell 1
       0000106012.00000     $ Remaining photons from carbon in cell 1
       0000026056.00102     $ Capture gammas from 56Fe in cell 1
       0000026056.00000     $ Remaining photons / gammas from 56Fe
       0000000000.00051     $ Remaining 1st level from (n,n') gammas
     10000000000.00000      $ Remaining gammas
```

2.4.4.12 LET: Tally Stopping Powers Instead of Energy

```
FTn LET
```

Linear energy transfer values provided in the energy bins of charged particle tallies are interpreted as stopping power values with units of MeV/cm.

2.4.4.13 ROC: Receiver-Operator Characterization

```
FTn ROC n
TFn  F D U S M C E T    F D U S M C E T
```

The FT ROC receiver-operator separates the tally into signal and noise and runs histories in batches of *n* source particles, where the total number of NPS source particles should be 50–100 times *n*. The 1st eight entries on the TFn tally fluctuation bin card point to the signal bin of the tally; the 2nd eight entries point to the noise bin. For example, the signal may be the value in one energy, time, cosine, or other bin, and the noise may be the value in some other bin. The probability of detection and the probability of false alarm are then printed in Table 163 of the MCNP output file.

2.4.4.14 PDS: Point Detector Sampling

```
FTn    PDS    c
```

Point detectors (n ending in 5, or F5 tallies) suffer convergence problems when collision sampling has a large variation in probabilities. For example, photon coherent scattering in the forward direction has orders of magnitude greater tally contribution than in the backward direction. In these cases—particularly with photon point detectors, F5:p—the PDS special treatment changes sampling to more often sample the rare high contribution events.

$c = 0$ or $c = -1$ Next-event estimator sampling is performed post-collision; only a single reaction and isotope is sampled (historic MCNP4 and MCNP5 behavior) (DEFAULT).

$c = 1$ Next-event estimator sampling is performed using post-collision sampling of the collision isotope and pre-collision sampling of all reaction channels (recommended for photons).

$c = 2$ Next-event estimator sampling is performed using pre-collision sampling of all collision isotopes and pre-collision sampling of all reaction channels.

2.4.4.15 FFT: First Fission Tally

```
FTn FFT [LKJI]
FUn   z₁   z₂   z₃    . . .
```

The first fission tally records the nuclide with the first fission or (n,xn) reaction. This special treatment was developed specifically for safeguards applications because the signal from fissions is roughly scaled by the number of fission neutrons emitted by the first fissioning nuclide. The optional LKJI values are

L = 0/1 Omit/include neutron and photon-induced fissions treated by model physics
K = 0/1 Omit/include neutron spontaneous fissions (PAR = SF source particles)
J = 0/1 Omit/include photon-induced fissions treated by library physics
I = 0/1 Omit/include neutron-induced fissions treated by library physics (E $<\sim20$ MeV)

The z_i values are ZAID identifiers, 0 for everything else, 16 for (n,xn), and 18 for some other fission. For example,

```
FTn FFT 0001
FUn 92238 0 16 18 94241 92235 94239
```

The FFT 0001 is the default, including only fissions treated by library physics. The tally is subdivided into user bins for those particles having had their first multiplicity reaction from ^{235}U fission (92235), ^{238}U fission (92238), ^{239}Pu fission (94239), ^{241}Pu fission (94241), any other fission (18), (n,xn) (16), or no multiplicity reactions at all before scoring.

Note that the FFT special treatment causes cell flagging (CF) and tally segmenting (FS) to apply at the location where the first fission or (n,xn) event occurs and not where the tally score is made. The FFT tally does not apply to F8 tallies.

2.4.4.16   COM: Compton Image Tally

```
FT8   COM T  A
```

The FT8 COM tally option produces a Compton image stored in an associated FIR radiography tally T using algorithm A (only A = 1 is currently available). The Compton image is formed from an FT8 PHL specification of dual-region coincidences of planar lattice tallies. At the end of each particle history, Compton/photoelectric energy deposition in the front/back of these dual-panel detectors is used to create a circular "image" of the incident photon on a specified image plane. For more details, see the MCNP code manual section 3.3.5.18.

2.4.4.17   CAP: Coincidence Capture

The coincidence capture tally is a key safeguards application that models multiplicity counting methods. The tally produces the factorial moments of the neutron multiplicity distribution. The use of this tally is described in Sect. 3.2. The FT8 CAP syntax is

$$\text{FT8 CAP } \left[-m_c\right]\left[-m_o\right] i_1 i_2 \ldots i_n \left[\text{GATE } t_d \ t_w\right] \left[\text{EDEP } t_g \ t_t\right]$$

where

m_c = optional maximum number of captures (Default = 21),
m_o = optional maximum number of moments (Default = 12),
i_n = capture nuclides such as 3006 or 5010 for 6Li or 10B,

t_d = predelay time (shakes),
t_w = gate width (shakes),
t_g = trigger tally number, and
t_t = trigger tally threshold (MeV) (Default=0.0).

Coincidence and multiplicity calculations are described in greater detail in Sect. 3.1. The FT8 tallies from Example 3.2 are

```
fc8 Ungated Coincidence tally
f8:n (10 11 12 13 14 15 16 17 18 19 20 21 22 23 24 25 26 27)
ft8   cap 2003
c
fc18 Gated Coincidence tally
f18:n (10 11 12 13 14 15 16 17 18 19 20 21 22 23 24 25 26 27)
ft18  cap 2003 gate 450 6400 $ 1us=10 shakes --> 4.5 us predelay
and 64 us gate
c
FC108 Full gate to compare moments with ungated F8
F108:n (10 16I 27)
FT108   CAP  2003 gate 0 1e20
c
FC116  Energy deposition tally to demonstrate FT118 CAP EDEP
F116:n (10 16I 27)
c
FC118  Energy deposition capture tally
F118:n  (10 16I 27)
FT118   CAP  EDEP 116 .5
```

The FT8 CAP tally counts the number of captures and records the multiplicity and calculates the moments. The data are stored in cosine bins that enable tally plotting because there are no multiplicity bins in the 8-dimensional tally array. In the standard tally output (in the cosine format), these data are hard to read, so they are also presented in more readable form in Print Table 118:

```
1 neutron captures, moments and multiplicity distributions.  tally       8         print
table 118

  weight normalization by source fission neutrons =        2154543
          cell  a is (10 11 12 13 14 15 16 17 18 19 20 21 22 23 24 25 26 27)

  cell:     a

  neutron captures on 3he
```

captures		histories	captures by number	captures by weight	multiplicity fractions by number	by weight	error
captures =	0	649116	0	0.00000E+00	6.49116E-01	3.01278E-01	0.0007
captures =	1	280151	280151	1.30028E-01	2.80151E-01	1.30028E-01	0.0016
captures =	2	58825	117650	5.46055E-02	5.88250E-02	2.73028E-02	0.0040
captures =	3	9573	28719	1.33295E-02	9.57300E-03	4.44317E-03	0.0102
captures =	4	1811	7244	3.36220E-03	1.81100E-03	8.40549E-04	0.0235
captures =	5	380	1900	8.81858E-04	3.80000E-04	1.76372E-04	0.0513
captures =	6	108	648	3.00760E-04	1.08000E-04	5.01266E-05	0.0962
captures =	7	29	203	9.42195E-05	2.90000E-05	1.34599E-05	0.1857
captures =	8	5	40	1.85654E-05	5.00000E-06	2.32068E-06	0.4472
captures =	9	2	18	8.35444E-06	2.00000E-06	9.28271E-07	0.7071
total		1000000	436573	2.02629E-01	1.00000E+00	4.64136E-01	0.0015

factorial moments	by number		by weight	
3he	4.36573E-01	0.0015	2.02629E-01	0.0015
3he(3he-1)/2!	1.04651E-01	0.0051	4.85722E-02	0.0051
3he(3he-1)(3he-2)/3!	2.42400E-02	0.0177	1.12506E-02	0.0177
3he(3he-1) .... (3he-3)/4!	6.94800E-03	0.0514	3.22481E-03	0.0514
3he(3he-1) .... (3he-4)/5!	2.16900E-03	0.1171	1.00671E-03	0.1171
3he(3he-1) .... (3he-5)/6!	6.19000E-04	0.2259	2.87300E-04	0.2259
3he(3he-1) .... (3he-6)/7!	1.41000E-04	0.3846	6.54431E-05	0.3846
3he(3he-1) .... (3he-7)/8!	2.30000E-05	0.5619	1.06751E-05	0.5619
3he(3he-1) .... (3he-8)/9!	2.00000E-06	0.7071	9.28271E-07	0.7071

These results are slightly different (though statistically the same) from those in Fig. 3.11, which is part of the more detailed discussion of coincidence counting in Sect. 3.2.6.2. The difference arises from running the problem on different computer platforms and operating systems. Tally F8 is the simplest form, counting all captures in ^{3}He in the cells listed on F8:n.

```
1 neutron captures, moments and multiplicity distributions.  tally      18          print
table 118

weight normalization by source fission neutrons =        2154543
            cell  a is (10 11 12 13 14 15 16 17 18 19 20 21 22 23 24 25 26 27)

cell:     a

neutron captures on 3he

time gate:  predelay = 4.5000E+02      gate width =  6.4000E+03
```

| pulses | | occurrences | occurrences | | pulse fraction | |
in gate	histogram	by number	by weight	by number	by weight	error
captures = 0	372500	0	0.00000E+00	3.72500E-01	1.72890E-01	0.0014
captures = 1	55609	55609	2.58101E-02	5.56090E-02	2.58101E-02	0.0044
captures = 2	7100	14200	6.59072E-03	7.10000E-03	3.29536E-03	0.0129
captures = 3	1134	3402	1.57899E-03	1.13400E-03	5.26330E-04	0.0332
captures = 4	187	748	3.47173E-04	1.87000E-04	8.67933E-05	0.0827
captures = 5	36	180	8.35444E-05	3.60000E-05	1.67089E-05	0.1924
captures = 6	7	42	1.94937E-05	7.00000E-06	3.24895E-06	0.5533
total	436573	74181	3.44300E-02	4.36573E-01	2.02629E-01	0.0045

factorial moments	by number	by weight
n	7.41810E-02 0.0055	3.44300E-02 0.0054
n(n-1)/2!	1.20890E-02 0.0201	5.61093E-03 0.0201
n(n-1)(n-2)/3!	2.38200E-03 0.0612	1.10557E-03 0.0612
n(n-1)(n-2) ... (n-3)/4!	4.72000E-04 0.1549	2.19072E-04 0.1549
n(n-1)(n-2) ... (n-4)/5!	7.80000E-05 0.3140	3.62026E-05 0.3140
n(n-1)(n-2) ... (n-5)/6!	7.00000E-06 0.5533	3.24895E-06 0.5533

Tally F18 is a gated capture tally. When a neutron is captured, a "gate" is opened. After a predelay of 450 shakes (4.5 μs), the number of captures is counted for the next 6400 shakes (64 μs).

```
1 neutron captures, moments and multiplicity distributions.  tally      108          print
table 118


weight normalization by source fission neutrons =        2154543
          cell   a is (10 11 12 13 14 15 16 17 18 19 20 21 22 23 24 25 26 27)


cell:      a


neutron captures on 3he


time gate:  predelay = 0.0000E+00     gate width = 1.0000E+20
```

| pulses | | occurrences | occurrences | | pulse fraction | |
in gate	histogram	by number	by weight	by number	by weight	error
captures = 0	350884	0	0.00000E+00	3.50884E-01	1.62858E-01	0.0014
captures = 1	70733	70733	3.28297E-02	7.07330E-02	3.28297E-02	0.0036
captures = 2	11908	23816	1.10539E-02	1.19080E-02	5.52693E-03	0.0091
captures = 3	2335	7005	3.25127E-03	2.33500E-03	1.08376E-03	0.0207
captures = 4	524	2096	9.72828E-04	5.24000E-04	2.43207E-04	0.0437
captures = 5	144	720	3.34178E-04	1.44000E-04	6.68355E-05	0.0833
captures = 6	36	216	1.00253E-04	3.60000E-05	1.67089E-05	0.1667
captures = 7	7	49	2.27426E-05	7.00000E-06	3.24895E-06	0.3780
captures = 8	2	16	7.42617E-06	2.00000E-06	9.28271E-07	0.7071
total	436573	104651	4.85722E-02	4.36573E-01	2.02629E-01	0.0036

factorial moments	by number		by weight	
n	1.04651E-01	0.0051	4.85722E-02	0.0049
n(n-1)/2!	2.42400E-02	0.0177	1.12506E-02	0.0177
n(n-1)(n-2)/3!	6.94800E-03	0.0514	3.22481E-03	0.0513
n(n-1)(n-2) ... (n-3)/4!	2.16900E-03	0.1171	1.00671E-03	0.1171
n(n-1)(n-2) ... (n-4)/5!	6.19000E-04	0.2259	2.87300E-04	0.2259
n(n-1)(n-2) ... (n-5)/6!	1.41000E-04	0.3846	6.54431E-05	0.3846
n(n-1)(n-2) ... (n-6)/7!	2.30000E-05	0.5619	1.06751E-05	0.5619
n(n-1)(n-2) ... (n-7)/8!	2.00000E-06	0.7071	9.28271E-07	0.7071

If the gate is infinite, as in F108, then the 1st, 2nd, 3rd, … moments are the same as the ungated 2nd, 3rd, 4th, … moments of the F8 tally.

```
1 neutron captures, moments and multiplicity distributions.  tally      118           print
table 118
```

weight normalization by source fission neutrons =   2154543
                    cell a is (10 11 12 13 14 15 16 17 18 19 20 21 22 23 24 25 26 27)

cell: a

Energy losses > 5.0000E-01 MeV in tally F116

	histories	captures by number	captures by weight	multiplicity fractions by number	by weight	error
captures = 0	791900	0	0.00000E+00	7.91900E-01	3.67549E-01	0.0005
captures = 1	183826	183826	8.53202E-02	1.83826E-01	8.53202E-02	0.0021
captures = 2	21834	43668	2.02679E-02	2.18340E-02	1.01339E-02	0.0067
captures = 3	2138	6414	2.97697E-03	2.13800E-03	9.92322E-04	0.0216
captures = 4	267	1068	4.95697E-04	2.67000E-04	1.23924E-04	0.0612
captures = 5	31	155	7.19410E-05	3.10000E-05	1.43882E-05	0.1796
captures = 6	3	18	8.35444E-06	3.00000E-06	1.39241E-06	0.5773
captures = 7	1	7	3.24895E-06	1.00000E-06	4.64136E-07	1.0000
total	1000000	235156	1.09144E-01	1.00000E+00	4.64136E-01	0.0021

factorial moments	by number		by weight	
n	2.35156E-01	0.0021	1.09144E-01	0.0021
n(n-1)/2!	3.02260E-02	0.0077	1.40290E-02	0.0077
n(n-1)(n-2)/3!	3.61100E-03	0.0302	1.67599E-03	0.0302
n(n-1)(n-2) ... (n-3)/4!	5.02000E-04	0.1080	2.32996E-04	0.1080
n(n-1)(n-2) ... (n-4)/5!	7.00000E-05	0.3440	3.24895E-05	0.3440
n(n-1)(n-2) ... (n-5)/6!	1.00000E-05	0.7211	4.64136E-06	0.7211
n(n-1)(n-2) ... (n-6)/7!	1.00000E-06	1.0000	4.64136E-07	1.0000

The F8 CAP tally scores when a capture occurs. This is a good approximation for detectors like ³He, in which almost 100% of captures lead to a detected pulse. The tally does, however, have the ability to count "captures" by energy deposition, which is useful for simulation of other detectors such as boron-lined counters. For example,

```
FT118   CAP  EDEP 116 .5
```

counts a capture for every track in the specified cells that deposit greater than 0.5 MeV of energy in cell 116. In the case of a ³He detector, a closer agreement with the F8 tally can be achieved by using the pulse-height energy deposition tally with a lower threshold to count all of the events:

```
FC128   *F8 energy deposition tally to demonstrate FT118 CAP EDEP
*F128:n (10 16I 27)
E128 100 NT
c
FC118   Energy deposition capture tally
F118:n  (10 16I 27)
FT118   CAP  EDEP 128 1e-10
```

The cells specified on the F118:n card are required but ignored; the tally is based on the energy deposition in tally F128. The *F8 heating tally is a surface estimator that tallies the difference between particles that enter and exit the cells specified by *F128. Note that a threshold of zero will not work because MCNP deducts 10^{-12} of the energy that passes through each cell in order to distinguish between particles that never entered the cell and those that entered but did not lose any energy. An example of the use of the EDEP option is given in Sect. 4.4.2.

2.4.4.18   PHL: Pulse-Height Light Tally

The pulse-height light special treatment, FT8, is described with the other F8 pulse-height tallies in Sect. 2.4.5.

2.4.5   Pulse-Height Tallies

Safeguards detectors do not measure flux or energy deposition directly. They measure pulses or signals. The MCNP pulse-height tally and its various forms are used to model scintillator detectors.

The F8 pulse height looks like just another cell tally

```
f8:n   c₁   c₂   c3   . . .
```

but it is unlike other MCNP tallies in that it does not model currents and fluxes or reaction rates that can be modeled with the Boltzmann transport equation; rather it is a non-linear tally. With the FT8 RES and FT8 CAP special treatments, it is a collision estimator that counts residuals and captures. In its simplest form, it is a surface tally, counting the energy that enters a cell and subtracting out what leaves to put a pulse of that energy in the tally energy bins.

The MCNP code's F1, F2, F4, F5, F6, and F7 tallies score the energy bins as the energy at which the event happened, not the amount deposited. Consequently, the E0 default energy can apply to either those standard tallies or pulse-height tallies, but not both. Charged particles, on average, emit the correct number of secondary particles. However, any given track may sample the production of more secondary particles that carry more energy than the incident particle which leads to a negative pulse-height energy deposition. Negative energy production from sampling too many secondary particles is nonphysical but correct, on average, to get the total energy deposited. Thus, pulse-height tallies should have a zero energy bin to tally these rare nonphysical imbalance events with negative scores. Particles that do not enter the pulse-height tally cell make no score. Particles that enter the pulse-height tally cell but deposit no energy need to be distinguished, so the MCNP code gives them an energy deposition of 10^{-12} times the particle's energy. Thus the energy bins for a pulse-height tally should be something like

```
e8    0 1e-6 .001 .01 .1 100ilog 101
```

The energy pulse is accumulated from all tracks of a particle's history, including those of its created secondary particles. Consequently, variance reduction—which affects the particle weight—greatly complicates the accumulation of these tracks. Variance reduction can be used with pulse-height tallies in some cases, but the algorithm is complicated and fragile. Thus, variance reduction should be avoided. If used, it should be done with caution [7].

A *F8 tally multiplies the pulse by the energy of the track at each collision in the pulse-height cell.

Consider a photon that enters a cell at 10 MeV, then exits at 9 MeV, then re-enters at 7 MeV, and exits at 6 MeV. The F6 tally would score 1 MeV in the 10 MeV bin and 1 MeV in the 7 MeV bin.

The *F8 tally would have 2 MeV in the 2 MeV bin. The F8 tally would have 1 pulse in the 2 MeV bin.

Pulse-height tallies may also be used for the FT8 PHL pulse-height light special treatment. The PHL treatment enables more realistic detector modeling including coincidence and anticoincidence. Example 2.20 has simple F8, energy *F8, and light FT8 PHL pulse-height tallies:

Example 2.20 NE213 Scintillator

```
NE213 scintillator
11  1 -0.9  -11      imp:n=1
20  0         -100 11 imp:n=1
22  0          100    imp:n=0


11  RCC 0 0 0 0 0 3.81 2.54
100 SO   100

sdef pos 0 0 -1 erg 15 vec 0 0 1 dir 1
m1    1001 1.213   6000 1
nps 1e5
print -85 -86 -162 -70 -98
prdmp 2j -1
mode n h
phys:n   J 20 J J J J 1 $ analog < 20 MeV, ion elastic recoil
cut:n 2J 0 0
cut:h j 0 0 0
FC6 Track length heating converted to light output
f6:h 11                  $ F6 is converted to light
DE6  LIN 0 0.2 0.4 0.6 0.8 1 1.2 1.4 1.6 1.8 2
           2.2 2.4 2.6 2.8 3 3.2 3.4 3.6 3.8 4
           4.2 4.4 4.6 4.8 5 5.2 5.4 5.6 5.8 6
           6.2 6.4 6.6 6.8 7 7.2 7.4 7.6 7.8 8
           8.2 8.4 8.6 8.8 9 9.2 9.4 9.6 9.8 10
DF6  LIN 0.00 0.01 0.04 0.08 0.12 0.17 0.23 0.29
           0.35 0.42 0.50 0.57 0.66 0.74 0.83 0.93
           1.02 1.12 1.23 1.34 1.45 1.56 1.67 1.79
           1.92 2.04 2.17 2.30 2.43 2.57 2.70 2.84
           2.99 3.13 3.28 3.43 3.58 3.74 3.90 4.05
           4.22 4.38 4.55 4.72 4.89 5.06 5.23 5.41
           5.59 5.77 5.95
FC16 Track length MeV/g deposited
f16:h   11                $ F16 is Mev/g deposited
FC26 Track length MeV deposited
f26:h   11
sd26   1   $ Divide by 1 rather than grams
e0    0.2 48I 10 11 13 15 17 20
FC8 Total light output
f8:h   11                 $ F8 is total light output
FT8    PHL 1 6 1 0
e8    0 1e-6 .001 .05 5I .35 .37 .4 9I 1.0 1.2 1.32 1.4 10I 3.6
      3.79 3.8 30I 10 6I 50
FC18 Total energy deposited
F18:h   11
Ft18   PHL 1 16 1 0       $ 0.37 1.32 and 3.79 are expt thresholds
```

```
e18   0 1e-6 .001 .05 5I .35 .37 .4 9I 1.0 1.2 1.32 1.4 10I 3.6
      3.79 3.8 30I 10 6I 50
FC28 Pulse height tally
f28:h  11
e28   0 1e-6 .001 .05 5I .35 .37 .4 9I 1.0 1.2 1.32 1.4 10I 3.6
      3.79 3.8 30I 10 6I 50
FC38 Pulse height energy deposition
*f38:h  11
e38   0 1e-6 .001 .05 5I .35 .37 .4 9I 1.0 1.2 1.32 1.4 10I 3.6
      3.79 3.8 30I 10 6I 50
```

Example 2.20 is an oversimplified NE-213 scintillator model that consists of a 3.81 cm high and 2.54 cm radius can with a mono-directional, 15 MeV neutron source directed up the z-axis of the can. This example is a coupled neutron-proton calculation: MODE N H. The 2nd PHYS:N entry turns on analog capture below 20 MeV. Analog capture is more evidently specified by setting the weight cutoffs, the 3rd and 4th entries of CUT: N 2J 0 0 to zero. The 7th PHYS:N entry turns on proton recoil when a neutron scatters off hydrogen, and the scattered hydrogen nucleus is transported as a proton. CUT:H J 0 0 0 turns on analog capture, and the 2nd entry lowers the proton energy cutoff to zero. The default for protons is a 1 MeV lower cutoff, but now the MCNP code will set it to 0.001 MeV (1 keV), the lowest possible for protons in the MCNP code.

The F16:H tally is a standard, simple, proton-heating or energy-deposition tally with units of MeV/g. The F26:H tally has a segment divisor, DS26, which overrides the divisor of grams, and thus it is the standard, simple, proton-heating or energy-deposition tally with units of MeV. The F6:H proton tally is converted to light output by the dose function, DE6 and DF6. The dose functions are obtained from experimental measurement as described in Chap. 8 of Ref. [8]. The tally is multiplied by the linearly interpolated point on DF6 at the linearly interpolated values between the energies on DE6. Because the tally is multiplied by as much as a factor of 6 at higher energies, the upper tally energy bin is exceeded, and the MCNP code issues the warning

```
warning. tally not scored beyond last energy bin.
nps =          176     tal =   8      erg = 5.2820E+01
```

The energies of the F6 tallies are set by the E0 default energy. The pulse-height energy bins must be specified for each pulse-height tally because pulse-height tallies and non-pulse-height tallies cannot both use the same E0 default energy binning.

Tally F28 is a simple pulse-height tally with units of pulses in each energy bin.

Tally *F38 is an energy pulse-height tally with units of MeV in each energy bin. The total bin value is 0.909 MeV, which agrees well with the F26 total bin of 0.907 MeV. F38 uses a surface estimator and F26 uses a track-length estimator.

Tallies F8 and F18 are pulse-height light tallies. The FT8 PHL syntax is

```
FT8    PHL n ta1 ba1 ta2 ba2 ... m tb1 bb1 tb2 bb2 ...
```

n = number of F6 tallies for the first detector region

tai, bai = pairings of tally number and F-bin number for the n F6 tallies of the first detector region

m = number of F6 tallies for the second detector region

tbi, bbi = pairings of tally number and F-bin number for the m F6 tallies of the second detector region

The pulse-height light tally gets its energy deposition in the cell not from the surface estimator weight balance or a collision estimator at each collision but rather from another energy deposition tally. The energy deposition tally may be a type 6 tally, as illustrated here (F6), or a type 8 tally (*F38). Because these heating tallies can have different particle types, the FT8 PHL tally can mix and match contributions from different particles. For example,

```
mode  n h d s t a
phys:n 6J 4
cut:h,d,t,s,a J 0 0 0
e0 100 nt
f116:h 3 4
f126:d 3 4
f136:t 3 4
f146:s 3 4
f156:a 3 4
fc208 Combined PHT for h,d,t,s,a
f208:n 3
e208 0 1e-3 99ilog 10 100
ft208 phl 5 116 1 126 1 136 1 146 1 156 1 0 GEB .01 .02 .04
```

In this example, the 7th PHYS:N entry enables neutrons that strike any of the light nuclei to generate secondary light ions from both elastic and selected inelastic reactions. Analog capture and the lowest possible energy cutoff are invoked with the CUT entries. The pulse-height light tally, FT8 PHL has five heating tallies in its one region, each using the energy deposition in the first cell, cell 3, of the F6 tally.

In addition to the F8 pulse-height tally, the *F8 energy pulse-height tally, and the FT8 PHL pulse-height light tally, there is the possibility of a coincidence–anticoincidence pulse-height light tally by having up to four FT8 PHL regions. For example,

```
f116:h 3 4
fc308 PHT for h in cell 3 no/yes having cell 4 energy deposition
f308:h (3 4)
e308 0 1e-3 99ilog 10 100
fu308 0 100
fq308 e u
ft308 phl 1 116 1 1 116 2 GEB .01 .02 .04
```

In this example, the FT308 PHL has 2 regions: tally F116 cell 3 (first cell) and tally F116 cell 4 (second cell). The first region, cell 3, uses energy bins E308. The second region, cell 4, uses user bins FU308 for energy bins. (A third region would use C cosine bins and a fourth region would use FS segment bins for energy.)

In this case (with two regions) the scores are made in a two-dimensional matrix, where one dimension represents the score in cell 4 (in user bin energies defined on the fu308 card) and the other dimension represents the score in cell 3 (in regular energy bins defined on e308 card). If there is no score in cell 4, the score recorded in cell 3 appears in the lowest energy bin of cell 4. If there is no score recorded in cell 3, the score in cell 4 appears in the lowest energy bin of cell 3. If there is no score in either cell there is no entry in the tally. The fq308 e u card causes the user bins (cell 4 score) to run horizontally and the regular energy bins (cell 3 score) to run vertically. The first row of data is the pulse height spectrum in cell 4 when there is no score in cell 3. The first column of data is the pulse height spectrum in cell 3 when there is no score in cell 4. The other bins show the energy relationship between coincident events.

Thus, coincidence/anticoincidence is achieved by seeing which parts of the pulse height were caused from contributions from which regions. As another example,

```
F8:N 5
FT8 PHL 1 6 1 1 16 1
E8 1.0 2.0 3.0 4.0 5.0 6.0 7.0 8.0
FU8 0 10
F6:E 5
F16:E 6
```

The FU8 0 10 card creates two user bins, one <0 and one 0-10. The first user bin of the pulse-height tally in cell 5 (F6 electron energy deposition) scores only when there is no energy deposition in cell 6 (F16); the second user bin records the pulse-height tally in cell 5 (from F6) only when there is energy deposition in cell 6 (F16) with $0 < E < 10$.

More information on the use of pulse height tallies can be found in [9].

2.4.6   Point Detectors and Next-Event Estimators

Point detectors enable the estimation of flux at a point. The probability of a neutron or a photon having a random walk exactly to a point is zero, but the flux at the point can be estimated using a next-event estimator. The next-event estimator may also be used to deterministically transport neutrons or photons to a spherical region—the basis of the variance-reduction technique DXTRAN (deterministic transport).

The next-event estimator accumulates the flux, Φ, at a point by a contribution of

$$\Phi = Wp(\mu)e^{-\sigma x} / 2\pi R^2$$

This estimate of Φ is accumulated at every source or collision event.

W = particle weight
$p(\mu) = p(\Omega)/2$ = scattering probability density function
R = distance between event and detector point
$e^{-\sigma x}$ = attenuation through all cells/materials along path R

Although the probability of scattering exactly in the direction, Ω, from the event to the detector is zero, the probability density function of the cosine, μ, assuming azimuthal symmetry, $p(\mu) = p(\Omega)/2$, is finite.

$$0 \le p(\mu) < \infty$$
$$-1 \le \mu \le 1$$

For photon coherent scattering with form factors, $p(\mu)$ is nearly a delta function at $\mu = 1$.

$$p(1) \approx \infty$$

Consequently, convergence is poor unless coherent scatter is turned off (PHYS:P 2J 1) or the special treatment, FT5 PDS, is applied to sample collision outcomes rather than incident reaction probabilities.

Convergence is also poor if there are large estimates from collisions close to the detector, in which case R approaches zero. The MCNP input for a point detector allows specification of a radius of exclusion, R_0. If the detector is in a vacuum, then no collisions will occur close to the detector, and R_0 can safely be set to zero. R_0 can also be set to zero if there are no large estimates from collisions close to the detector as is often the case in air. If there are large collisions close to the detector, then R_0 can be specified as about a mean free path; all collisions inside this radius assume a flat flux uniform distribution, which is a good approximation if R_0 is small. In our experience a better approach is to use the nested DXTRAN spheres variance reduction technique to better sample collisions near the point detector.

The attenuation term, $e^{-\sigma x}$, slows calculations so much that a Russian roulette variance reduction technique is automatically turned on to prevent prohibitive running times. This attenuation takes a lot of computer time because the transmission from the event to the detector requires transport through each intervening cell and material. The non-attenuation terms of the next-event estimator are known at the event before transmission to the detector

$$Wp(\mu)/2\pi R^2.$$

If these non-attenuation terms are less than a specified value, k, then Russian roulette is played on the transmission to the detector. That is, if

$$X = Wp(\mu)/2\pi R^2 < k,$$

then the transmission is rouletted with probability

```
1 - X / k
```

The transmission is continued with weight $W' = W\,k/X$ with probability X/k.

The value of k is specified on the DD input card along with a parameter, M, controlling diagnostic prints (MCNP manual section 3.3.6.11). The default for k is $k = 0.1\Phi$; that is, the default Russian roulette cutoff value is a tenth of the average value of all previous Φ estimates. This default usually results in one transmission to the detector for each source history. If $k = 0$, then no Russian roulette is played, and all source and collision events cause a time-consuming transmission to the detector. If $k < 0$, then Russian roulette is based on the constant k. If $k > 0$, then Russian roulette is based on the current value of $k\Phi$.

Whereas the value of Φ changes with each history, the Russian roulette—and hence, transport—are not identical sampling schemes, and the Central Limit Theorem of statistics is violated. Though a mathematician would argue that the estimated relative errors are then wrong, the effect is negligible. Nevertheless, it is highly recommended that a short calculation be made to determine the average value of Φ and then a tenth of this value be used as the negative value of k. Setting k negative no longer violates the Central Limit Theorem.

Point detectors are specified as

```
F5:p   X   Y   Z   R₀
```

X, Y, Z = point detector location coordinates
R_0 = radius of exclusion for collisions close to detector

The MCNP code also allows ring detectors, which are more efficient but useful only for geometries rotationally symmetric about a major axis. The code also allows radiography tallies, which use the point detector repeated over a grid to model radiographical images. Also the MCNP code uses the same next-event estimator point detector mechanics for DXTRAN spheres. These are illustrated in Example 2.22.

Example 2.21 illustrates many of the above issues.

Example 2.21 Point Detector Example Using Scintillator Geometry of Example 2.20

```
NE213 scintillator
11  1 -0.9  -11      imp:n=1
20  0       -100 11 imp:n=1
22  0        100     imp:n=0
```

```
11   RCC 0 0 0 0 0 3.81 2.54
100 SO   100

sdef pos 0 0 -1 erg .5 vec 0 0 1 dir 1 ARA=1
m1    1001 1.213   6000 1
nps 1e5
print -85 -86 -162 -70 -98
prdmp 2j -1
MODE p
PHYS:p 2J 1
FC15 Photon point detector
F15:P   0 0 50 0
DD15   -3e-7 500
```

Example 2.21 uses the scintillator geometry and materials of Example 2.20 and converts to a photon-only problem (MODE p). The source has been modified to be 500 keV, ERG = 0.5 MeV—in the range of many coherent scatters. Because the source is mono-directional (vec 0 0 1 dir 1), the ARA area parameter must be added, or the MCNP code terminates with the following fatal error:

```
fatal error.   sdef requires ara if dir=constant with detectors
or dxtran.
```

A point detector 51 cm in the z-direction from the source point at 0,0,-1 has been added.

```
        F15:P   0 0 50 0
```

If the PHYS:P and DD15 cards are omitted, then the coherent scattering causes unacceptable convergence with large relative errors. The first 600 large detector scores print a diagnostic. A large detector score is $> M\Phi$ or $> -Mk$, which is input on the DD card. These messages appear between the 1st 50 particle histories and the problem summary table in the output file. When the PHYS:P and DD cards are omitted in the problem of Example 2.21, the first two lines are

```
det    t            wgt          psc        amfp        ddetx
 1 4.1480E-01 1.0000E+00 5.8540E+03 3.4184E-02 4.6590E+01

  radius         erg      cell      nps  nch  p   nrn    ipsc
0.0000E+00  5.0000E-01     11      1126   1        7      6
```

where

det = 1, the 1st detector (F15) in the problem;
t = 4.1480E-01 = detector tally estimate;
wgt = 1.0 = W = particle weight at event;
psc = 5.8540E + 03 = p(μ) density function value >> 1.0;
amfp = σ_t*x = sum of macroscopic cross section times transmission distance
through each cell of geometry between event and detector;
ddetx = 4.6590E+01 = R = distance from event to detector;
radius = R_0 = exclusion radius for collisions close to detector;
erg = 5.0000E-01 = particle energy (500 keV);
cell = 11 = cell where event happened;
nps = 1126 = history number of this large score;
nch = 1 = number of collisions so far in this history;
p = particle type (photon);
nrn = 7 = random number in this history of the event; and
ipsc = 6 = scattering law of this collision (coherent).

When the PHYS:p 2J 1 is added to the input file, coherent photon scatter is turned off. The large score diagnostics for scores 1000 times the average do not appear because the largest scores are only 100 times the average. The average tally per history is 2.85337E-06. Convergence is acceptable.

The detector diagnostics card controls both the detector Russian roulette and the large scores diagnostics:

```
DD15   -3e-7 500
```

The DD card plays Russian roulette on all scores < 3.e−7, which was chosen to be 0.1 times the average score of 1.85337E−06 of the run without the DD card. The large score diagnostics are printed for any scores >500 × 3e−7 = 1.5e−4.

The tally 15 printout and diagnostics are

```
1tally       15         nps =       100000         Photon point detector
            tally type 5     particle flux at a point detector.             units    1/cm**2
            particle(s): photons


 detector located at x,y,z = 0.00000E+00 0.00000E+00 5.00000E+01
                2.84992E-06 0.0158


 detector located at x,y,z = 0.00000E+00 0.00000E+00 5.00000E+01
 uncollided photon flux without coherent scattering
                1.06682E-07 0.1217
```

detector score diagnostics		cumulative fraction of	tally per	cumulative fraction of
maximum score	transmissions	transmissions	history	total tally
3.00000E-07	19229	0.71919	6.00000E-12	0.00000

6.00000E-07	4	0.71934	1.60652E-11	0.00001
1.50000E-06	6	0.71956	6.06829E-11	0.00003
3.00000E-06	1	0.71960	1.65946E-11	0.00003
6.00000E-06	6	0.71983	2.57346E-10	0.00013
1.50000E-05	453	0.73677	6.36520E-08	0.02246
3.00000E-05	3851	0.88080	8.08683E-07	0.30622
3.00000E-04	3187	1.00000	1.97722E-06	1.00000
1.00000E+36	0	1.00000	0.00000E+00	1.00000

```
average tally per history = 2.84992E-06          largest score = 2.86963E-04
(largest score)/(average tally) = 1.00692E+02    nps of largest score =      53475
score contributions by cell
         cell      misses       hits    tally per history    weight per hit
    1     11        8707       26737    2.84992E-06          1.06591E-05
    2     20      100000           0    0.00000E+00          0.00000E+00
          total   108707       26737    2.84992E-06          1.06591E-05


score misses
    russian roulette on pd                     0
    psc=0.                                103806
    russian roulette in transmission        4798
    underflow in transmission                103
    hit a zero-importance cell                 0
    energy cutoff                              0
```

Note that only 0.00013 of the total score comes from the 71.983% of the transmissions with a score less than 6e−6. Thus, a better DD choice would be

```
DD15   -6e-6 20
```

Whereas the largest score is 2.86963E−04, it is useful to have the large score diagnostics for scores larger than 20 × 6e−6 = 1.2e−4. The largest/average score is 100, which is good. If this ratio is much larger, there are usually convergence problems.

Whereas the Russian roulette criteria are identical for all histories, the history of the largest score, nps = 53475, can be rerun to see what made the score large. If the default DD is used, then the criteria for the Russian roulette change for each history and are unknown and thus histories cannot be rerun.

Note that "score contributions by cell" can be used to play Russian roulette on the contributions from different cells using a PD card; cells with a low weight per hit often should have PD values <1.

One might ask how having each source or collision event score directly to the point detector gives a believable flux estimate. The next-event estimator is robust and accurate and any claim to the contrary is probably user error. Any score that is made in any tally comes from the last event preceding it. The next-event estimator is justified by calculating the next event from a source or collision that is the last event before making a score.

2.5   Plotting

The MCNP code supports plotting with the X Windows graphics system. The window manager on the display device must support the X protocol with software such as Xming. (Linux supports the X protocol natively).

The MCNP code makes geometry, cross-section, and tally plots. The geometry plots can be interactive or command driven. The cross-section and tally plots are command driven. Mesh tally plots overlay a mesh of tallies over the geometry plot. All MCNP plots can be made from the RUNTPE continuation file or during the execution of a problem. Tally plots can also be made from MCTAL files, which are created by the PRDMP 3rd entry. (See MCNP Manual [2] section 3.3.7.2.3).

2.5.1   Geometry Plotting and Command Files

All MCNP plots can use command files. Whenever an the MCNP code is run in the plotting mode, a COMOUT file (with an appended c if the NAME option is used) is created. This file can then be used or edited to make a series of plots. For example, consider the MCNP input file "ex11_6a" of Example 2.6. The geometry plots of Fig. 2.9 through Fig. 2.13 through can be reproduced with the following MCNP execution command:

```
MCNP6  i=ex11_6a  n=plot11_6a.  com=com11_6a ip
```

The command file, "com11_6a" is

```
py 6 or 10 6 0 ex 20 la 0 0 pause
lev 0 la 2 2 cel mbody off pause
lev 1 la 0 2 cel pause
lev 2 la 0 0 pause
OR 10 6 -5 PY 6 EX 7 LEV 2 LA 0 1 IJK pause
pause
end
end
```

All plot commands can be shortened to two or three characters (until non-ambiguous). The commands PY, ORigin, Extent, Label, and LEVel are explained in the MCNP manual and also listed in the plot help package by typing "?", "help," or "option" when in the interactive plotter. The PAUSE command stops the plot and sends the following message to the plot terminal:

```
strike any key to continue. quit returns to interactive plot.
```

Striking any key (we prefer "enter") makes the next picture up to the next pause command. Typing "quit" returns to the interactive plot mode.

For geometry plots, the help command lists only possible commands; the same plots may be made interactively. For cross-section and tally plots, the MCPLOT help command provides a complete description of each command and its options.

For geometry plots, the command file commands may be entered in the command window, in the lower right of the plot window where it says "click here or picture or menu," or by clicking the interactive buttons. The equivalent interactive buttons for the com11_6a file are: ZX, OR, ZOOM 5, L1; then LEV, MBODY, L1, L2; LEV, L1; then LEV, L2; then IJK (on right margin), L2, ZOOM 3, OR.

Note that the NAME=plot11_6a command will cause a new COMOUT command file, plot11_6a.c, to be made. The COMOUT file (or .c file) can then be edited to reproduce or make even fancier plots with labels, titles, scales, etc.

While the MCNP code is running transport, the geometry plotter can be called by <ctrl-c> M to get the MCPLOT tally plot and then "PLOT" to get the geometry plotter. The PAC and N buttons (lower right), followed by L2 for cell quantities, can be used to overlay the geometry with the number of particles that enter, population, number of collisions, etc., of each cell in the geometry plot. N refers to the column number in the MCNP output file particle activity by cell (PAC) printout.

The command "LA 2 2" doubles the size of cell and surface labels. Other geometry plot buttons are

- UP, RT, DN, LF (on top menu) to go up, right, down, or left one frame;
- REDRAW on the bottom menu is needed to refresh the picture when the refresh is not done automatically;
- CURSOR to enclose a region on which to zoom in;
- CellLine to not outline cells by drawing surfaces;
- COLOR to toggle surface-only plots without color and to color by other cell quantities first clicked on the right side buttons. For example, "VOL" then "COLOR" will make the cells colored proportionately to cell volume;
- SCALES to get a grid of dimensions in centimeters;
- LEVEL to toggle repeated structures levels;
- XY, YZ, ZX to get PZ, PX, and PY plots;
- L1, L2 to display surface and cell labels;
- MBODY to toggle macrobody facets on and off; and
- LEGEND to get color scaling.

If the plot is moved or otherwise goes blank, clicking in the upper left corner will restore the plot. When entering commands from the terminal, the command "INTERACT" will return to the interactive plot mode. In the interactive plot mode, clicking the "PLOT>" button on the bottom returns to the command mode. Terminating the plotting session should be done by clicking "END" on the bottom menu because closing the plot window will cause the MCNP code to crash and lose files.

One way to print a plot to another document without using the interactive buttons, first go to the command mode on the terminal. If in the interactive mode, click "PLOT>" on the bottom menu. From the command mode, enter the command

"VIEWPORT SQUARE" and only the square plot will appear. Click inside the picture and then take a screenshot (<ctrl-alt><print screen> on Windows systems) and then click inside the document <ctrl-v> (on Windows). To exit the command prompt mode and return to the interactive plotter, enter "INTERACT." Alternatively the "file" command can be used to produce a postscript format file.

2.5.2   Cross-Section Plotting

MCNP cross sections can be plotted for neutrons, photons, photonuclear data, protons, and other cross-section libraries in ACE (A Compact ENDF) format [10]. MCNP stopping powers for electrons can also be plotted. The cross-section plotter is command driven, not interactive. The MCNP code is run in the IXZ mode.

To reproduce all the figures in Sect. 2.2, "Materials and Cross Sections," the MCNP code is run as

```
MCNP6   IXZ   i=EX152   N=J152.   COM=COM152
```

In this example, EX152 is the input file of Example 2.11 with MODE P E changed to MODE N P E. The MCNP code will produce files with names J152.o, J152.c, etc. The plot command file is COM152:

```
xs m313 cop xs 92238.80c cop xs 92235.80c
pause
xlim 1e-11 1e-5 xs 6000.80c mt 1 cop xs grph.10t mt 1 cop mt 2
cop mt 4
pause
xlim .001 100 ylim .01 1e7 par p mt -5 xs 1000.04p cop xs 8000.04p &
cop xs 40000.04p cop xs 92000.04p
pause
ylim .01 1e7 par p xs 92000.04p mt -5 cop mt -1 cop mt -2 cop
mt -3 cop mt -4
pause
ylim 1 200 par e xs m609 cop xs m201 cop xs m313
pause
par e xs m313 ylim .5 30 mt 3 cop mt 1 cop mt 2
pause
end
end
```

The cross section plot commands are largely the same as for the tally plotter (see MCNP manual [2] section 5.3). The commands relating to the picture such as XLIM, YLIM, etc., are the same; only commands unique to tallies or data are different. Once a command is set, it holds for all subsequent plots until specifically changed. RESET ALL resets to the default commands. XS, PAR, and MT are specific for the cross-section plotter. PAR defaults to neutrons, and if neutrons are not in the MCNP input file, nothing will be plotted until PAR is set to a particle type on the MODE card of the input file. All the possible plot and tally commands are listed when "?", "help," or "options" is entered. Each command is described in detail by entering the command name after the "?", "help," or "options." For example, "help XS" prints on the terminal:

```
xs      > Syntax: xs m
             Plot a cross-section according to the value of m.
              Option 1: m=Mn, a material card in the INP file.
           Example: XS M15. The available materials will be listed
                if a material is requested that does not exist in the
                INP file.
         Option 2: m=z, a nuclide ZAID. Example: XS 92235.50c. The
              full ZAID must be provided. The available nuclides will
              will be listed if a nuclide is requested that does not
                exist in the INP file.
             See "mt", "par".
```

"Help MT" describes how to get specific reactions in the cross-section plotter:

```
mt      > Syntax: mt m
             Plot reaction n of material XS m. The default is the
             total cross section. The available reaction numbers
             will be listed if one enters a reaction number that
             does not exist (e.g., 999).
             See "par", "xs".
```

Note that only cross sections and reactions specified in the MCNP input file may be plotted and they must be fully specified. XS=1000 fails; XS=1000.04p is required in this example. If a cross section is requested that is not available, then the plotter will list the ones available:

```
xs 100
   can't find zaid        100   available zaids:
                     1001.80c
                     6000.80c
                     7014.80c
                     8016.80c
                    40090.80c
                    40091.80c
                    40092.80c
                    40094.80c
                    40096.80c
```

```
                         92235.80c
                         92238.80c
                          benz.10t
                          grph.10t
                          lwtr.10t
                          poly.10t
                         1000.04p
                         6000.04p
                         7000.04p
                         8000.04p
                        40000.04p
                        92000.04p
```

If electrons have been loaded, they will have a similar ZAID set as the photons, with p changed to e. (The electron libraries are 03e and 01e.)

Similarly, when plotting a nuclide like xs=92235.80c (but not a material like xs=m311), specifying an unavailable MT reaction number causes the MCNP cross-section plotter to print the available reactions:

```
mt -12
available (not expunged) endf reactions:
 for isotope   92235.80c
 zaid   92235.80c has :
        mt = 16.
        mt = 17.
        mt = 18.
        mt = 37.
        mt = 51.
        mt = 52.
        mt = 53.
 . . .
        mt = 89.
        mt = 90.
        mt = 91.
        mt =102.
        mt =  4.
 some mts may have been expunged or may not exist
 in the material or zaid cross sections given.
 if an mt exists but is not listed here, call it out
 with an fm tally card.
```

The principal reaction MT values are shown in Table 2.1.

The S(α,β) reactions are 1 = total, 2 = elastic, 4 = inelastic.

The neutron cross sections do not include thermal effects. The photon cross sections do not include form factors for coherent and incoherent scattering.

In command file COM152, the other commands are:

COPlot = overlay the next plot on the present one to plot multiple curves;
PAUSE = do not advance to the next plot until a key is struck. If a number, N, fol-
    lows, then hold the plot for N seconds;
XLIMs = set the range of x-axis values; and
YLIMs = set the range of y-axis values.

Model cross sections cannot be plotted. Secondary energy and angle distributions cannot be plotted.

2.5.3 Tally Plotting

All MCNP tallies, most KCODE criticality quantities, and statistical information
can be plotted with the MCNP tally plotter. The plots can be made during the course
of a calculation either by interrupt, <ctrl-c><m>, or periodically with the MPLOT
run-time plotter (MCNP manual [2] section 3.3.7.2.5), which sends a tally plot to
the terminal after specified numbers of histories or KCODE cycles. After a calculation, the RUNTPE file can be used to produce geometry, cross section, or tally plots
and the MCTAL file can be used for all tally plots. These capabilities also apply to
MESH tallies described in the next section.

    RUNTPE continue files can be read only by the version and operating system
that wrote them. MCTAL files are ASCII files in a backward-compatible format, so
MCTAL files written by different versions of the MCNP code on different operating
systems many years ago may still be used to plot tallies with the latest MCNP version. Furthermore, MCTAL files are much shorter and easier to store. Different
MCNP runs may be plotted against each other from any combination of MCTAL
and RUNTPE files.

    Tally plots are made with the command

```
MCNP6   Z
```

after which the MCNP code will ask for a RUNTPE file name (RUNtpe =) or an
MCTAL file name (RMCtal =).

    The following plot command file, COM153, plots all tallies of Example 2.20 and
Example 2.21 with MCTAL files j142.m and j143.m:

```
rmc=j142.m tal 6 xlim .2 20 loglog la "light output" &
title 1 "Proton Heating Tally with DE/DF dose function"
pause
tal 26 xlim 0 10 ylim .03 .2 linlog la "energy deposition" &
Title 1 "Proton Energy Deposition"
```

```
pause
tal 8 xlim .01 10 ylim .001 .2 linlog la "light output pulses" &
title 1 "Based upon light output F6: FT8  PHL  1 6 1"
pause
tal 18 ylim .005 .04 la "energy deposition pulses" &
title 1 "Based upon energy deposition F16: FT8 PHL  1 16 1"
pause
tal 28 xlim 0 10 ylim .005 .03 title 1 "Simple pulse-height
F8 tally"
pause
tal 18 Title 1 "Comparison of F8 and FT8 PHL 16" la "FT8 PHL
1 16 1" &
cop tal 28 la "Simple F8"
pause
tal 38 ylim .001 .1 title 1 "*F8 Energy-Deposition" la "energy
deposition"
pause
reset all
rmc j143.m tfc m title 1 "TFC M --- Mean Value Convergence"
la "mean"
pause
end
end
```

Figure 2.25 shows the plot of the F6 tally in Example 2.20:
The F6 tally of Example 2.20 is repeated here to accompany the plot of Fig. 2.26.

```
FC6 Track length heating converted to light output
f6:h 11                       $ F6 is converted to light
DE6  LIN 0 0.2 0.4 0.6 0.8 1 1.2 1.4 1.6 1.8 2
           2.2 2.4 2.6 2.8 3 3.2 3.4 3.6 3.8 4
           4.2 4.4 4.6 4.8 5 5.2 5.4 5.6 5.8 6
           6.2 6.4 6.6 6.8 7 7.2 7.4 7.6 7.8 8
           8.2 8.4 8.6 8.8 9 9.2 9.4 9.6 9.8 10
DF6  LIN 0.00 0.01 0.04 0.08 0.12 0.17 0.23 0.29
           0.35 0.42 0.50 0.57 0.66 0.74 0.83 0.93
           1.02 1.12 1.23 1.34 1.45 1.56 1.67 1.79
           1.92 2.04 2.17 2.30 2.43 2.57 2.70 2.84
           2.99 3.13 3.28 3.43 3.58 3.74 3.90 4.05
           4.22 4.38 4.55 4.72 4.89 5.06 5.23 5.41
           5.59 5.77 5.95
e0   0.2 48I 10 11 13 15 17 20
```

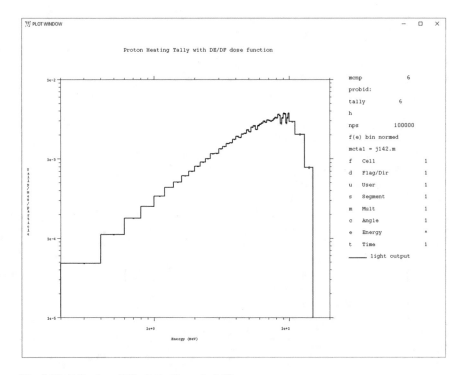

Fig. 2.25 Tally plot of F6 tally in Example 2.20

The tally plot commands are

```
rmc=j142.m tal 6 xlim .2 20 loglog la "light output" &
title 1 "Proton Heating Tally with DE/DF dose function"
```

Figure 2.26 shows the plot of the F26 tally in Example 2.20.

```
FC26 Track length MeV deposited
f26:h   11
sd26   1   $ Divide by 1 rather than grams
e0   0.2 48I 10 11 13 15 17 20
```

The tally plot commands are

```
rmc=j142.m tal 26 xlim 0 10 ylim .03 .2 linlog  &
la "energy deposition" Title 1 "Proton Energy Deposition"
```

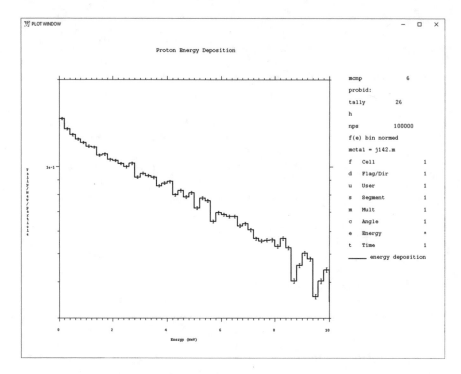

Fig. 2.26 Tally plot of F26 tally in Example 2.20. Note how the eight tally dimensions—F, D, U, S, M, C, E, T—are listed on the right

Figure 2.27 shows the plot of the F8 tally in Example 2.20.

```
FC8 Total light output
f8:h    11                    $ F8 is total light output
FT8     PHL 1 6 1 0
e8     0 1e-6 .001 .05 5I .35 .37 .4 9I 1.0 1.2 1.32 1.4 10I 3.6
       3.79 3.8 30I 10 6I 50
```

Note that the energy deposition for this F8 tally is from the heating tally F6 and does not require energy bins for the F6 tally.

The tally plot commands are

```
rmc=j142.m tal 8 xlim .01 10 ylim .001 .2 linlog &
la "light output pulses" &
title 1 "Based upon light output F6: FT8   PHL  1 6 1"
```

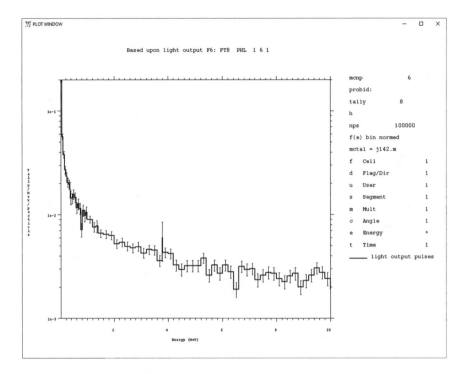

Fig. 2.27 Tally plot of F8 tally in Example 2.20. Note how the eight tally dimensions—F, D, U, S, M, C, E, T—are listed on the right

Figure 2.28 shows the plot of the F18 tally in Example 2.20.

```
FC18 Total energy deposited
F18:h   11
Ft18   PHL 1 16 1 0      $ 0.37 1.32 and 3.79 are expt thresholds
e18   0 1e-6 .001 .05 5I .35 .37 .4 9I 1.0 1.2 1.32 1.4 10I 3.6
      3.79 3.8 30I 10 6I 50
```

Note that the energy deposition for this F8 tally is from the heating tally F16 and does not require energy bins for the F16 tally.

The tally plot commands are

```
rmc=j142.m tal 18 ylim .005 .04 la "energy deposition pulses" &
title 1 "Based upon energy deposition F16: FT8 PHL  1 16 1"
```

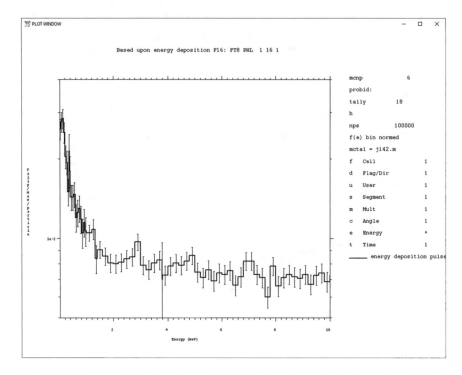

Fig. 2.28 Tally plot of F18 tally in Example 2.20. Note how the eight tally dimensions—F, D, U, S, M, C, E, T—are listed on the right

Figure 2.29 shows the plot of the simple (no FT special treatment) F28 tally in Example 2.20.

```
FC28 Pulse height tally
f28:h   11
e28   0 1e-6 .001 .05 5I .35 .37 .4 9I 1.0 1.2 1.32 1.4 10I 3.6
      3.79 3.8 30I 10 6I 50
```

Note that the energy deposition for this F8 tally is from a surface/collision estimator, which adds the energy entering the cell and produced by secondary particles in the cell and then subtracts out the energy of particles leaving the cell.

The tally plot commands are

```
rmc=j142.m tal 28 xlim 0 10 ylim .005 .03 &
title 1 "Simple pulse-height F8 tally"
```

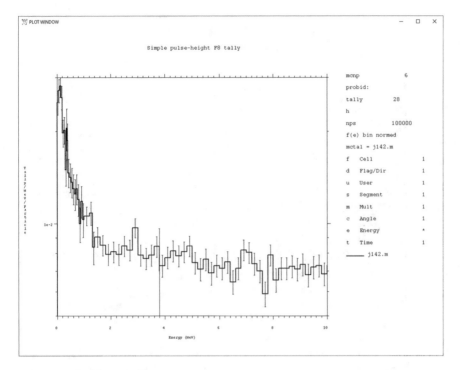

Fig. 2.29 Tally plot of F28 tally in Example 2.20. Note how the eight tally dimensions—F, D, U, S, M, C, E, T—are listed on the right

The name of the file is "j142.m," the name of the MCTAL file being read. The LAbel command should be used to get a better label.

Figure 2.30 shows a coplot of tallies F18 and F28 and demonstrates the capability to plot multiple curves with the MCNP plotter.

```
FC18 Total energy deposited
F18:h   11
Ft18   PHL 1 16 1 0        $ 0.37 1.32 and 3.79 are expt thresholds
e18   0 1e-6 .001 .05 5I .35 .37 .4 9I 1.0 1.2 1.32 1.4 10I 3.6
      3.79 3.8 30I 10 6I 50
FC28 Pulse height tally
f28:h   11
e28   0 1e-6 .001 .05 5I .35 .37 .4 9I 1.0 1.2 1.32 1.4 10I 3.6
      3.79 3.8 30I 10 6I 50
```

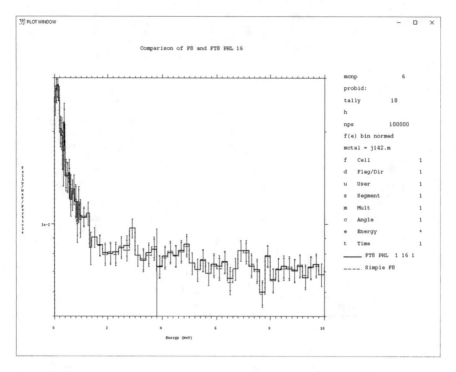

Fig. 2.30 Plot of Example 2.20 F18 and F28 tallies plotted together. Note how the eight tally dimensions—F, D, U, S, M, C, E, T—are listed on the right

Note that the pulses of the F18 pulse-height light tally energy deposition are from the F6 track length heating tally. The pulses of the F28 simple pulse-height tally are from the energy balance of surface and collision estimators. The two tallies agree within statistics despite being generated from totally independent estimators.

The tally plot commands are

```
tal 18 Title 1 "Comparison of F8 and FT8 PHL 16" &
la "FT8 PHL  1 16 1" cop tal 28 la "Simple F8"
```

Figure 2.31 shows tally *F38, an energy deposition pulse-height tally:

```
FC38 Pulse height energy deposition
*f38:h  11
e38  0 1e-6 .001 .05 5I .35 .37 .4 9I 1.0 1.2 1.32 1.4 10I 3.6
     3.79 3.8 30I 10 6I 50
```

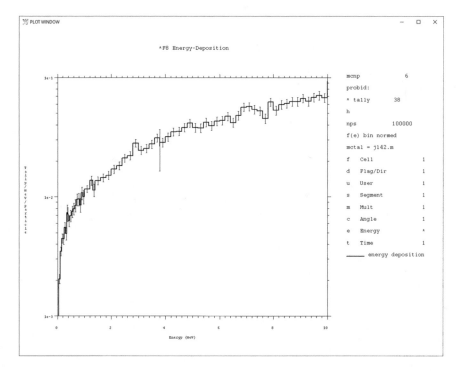

Fig. 2.31 Plot of Example 2.20 *F38 energy deposition pulse-height tally. Note how the eight tally dimensions—F, D, U, S, M, C, E, T—are listed on the right

The total tally bin is 0.908 ± .01 MeV, which is the same (within statistical uncertainty) energy deposition, 0.905 ± .01 MeV, as the F26 track length tally with the SD divisor to get units of MeV:

```
FC26 Track length MeV deposited
f26:h  11
sd26   1   $ Divide by 1 rather than grams
```

The tally plot commands are

```
rmc=j142.m tal 38 xlim 0 10 ylim .001 .1 &
title 1 "*F8 Energy-Deposition" la "energy deposition"
```

Figure 2.32 shows how it is possible to plot tallies with no energy, time, cosine, etc., bins—that is, tallies with a single number as the total tally.

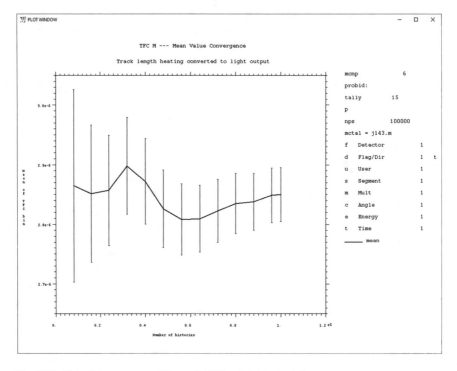

Fig. 2.32 Plot of convergence of Example 2.21 point detector tally

It also shows tally convergence of the point detector tally in Example 2.21:

```
FC15 Photon point detector
F15:P  0 0 50 0
DD15  -3e-7 500
```

The convergence and all convergence criteria of the 10 statistical tests can be plotted with the TFC plot command:

```
rmc  j143.m  tfc  m  title  1  "TFC  M  ---  Mean  Value  Convergence"
la "mean"
```

Note that because "TITLE 2" is not set, the second title line is carried forward from a previous plot.

Figure 2.33 shows the KCODE criticality plot of the principal estimators vs the cycle number using the MCNP tally plotter for the problem of Example 2.14.

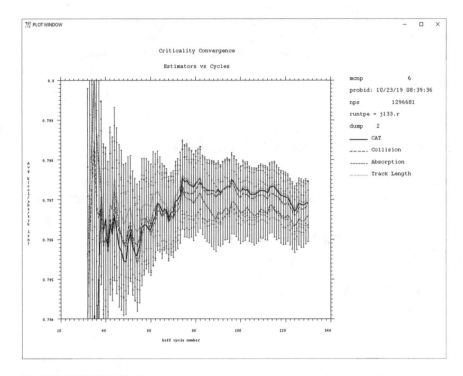

Fig. 2.33 KCODE Criticality plot

The plot COM file is

```
run=j133.r ylim .794 .8 kc 16 la "CAT" &
title 1 "Criticality Convergence" title 2 "Estimators vs Cycles" &
cop kc 11 la "Collision" cop kc 12 la "Absorption" &
cop kc 13 la "Track Length"
```

The various KCODE commands are listed when running the MCNP code in the tally plot mode and entering "HELP KCODE." Here, the multiple curves KC = 16, 11, 12, and 13 are plotted together. KC = 16 is the cumulated covariance-weighted collision-absorption-track length (CAT) estimator of k_{eff}; KC = 11 is the cumulative collision estimator of k_{eff}; KC = 12 is the cumulative absorption estimator of k_{eff}; and KC = 13 is the track length estimate of k_{eff}. Because Example 2.14 did not have a PRDMP 3rd entry to create a MCPLOT file, the RUNTPE file, RUN=j133.r, has to be used.

Run-Time Plotting: MCNP plots may also be generated during the course of a run with the MPLOT input card. For Example 2.14, add the following to the MCNP input file:

```
MPLOT freq 5 kc 16 cop kc 11 cop kc 12 cop kc 13
```

FREQ=5 puts up a plot of the calculation every 5 KCODE cycles. For non-KCODE calculations, FREQ is the number of histories between which plots will be sent to the computer terminal. MPLOT arguments are MCPLOT commands. A user can visualize source convergence with an FMESH plot (see next section) called on MPLOT, for example.

2.5.4 Mesh, Radiography, and Ring Tallies

Example 2.22 illustrates a number of tally capabilities. It also uses the vertical input format for the source definition with the # operator (MCNP manual [2] section 2.8.2).

Example 2.22 NE-213 Detector with Point, Ring, Radiography, and Mesh Tallies

```
NE213 scintillator
11   1 -0.9  -11      imp:n=1
20   0        -100 11 imp:n=1
22   0         100    imp:n=0

11  RCC 0 0 0 0 3.81 2.54
100 SO   100

MODE n h
m1    1001 1.213  6000 1
mx1:h   J         6012
nps 1e5
print -85 -86 -162 -70 -98
prdmp 2j 1
c
SDEF PAR N x=d21 y=d22 z -50 erg 100 vec 0 0 1 dir 1
     ARA=56
#     SI21   SP21  SI22   SP22
        -4     0   -3.5     0
         4     1    3.5     1
c
TALNP  55 65
FC15:n Point detector at side
F15:n  4 0 2  0
dd15 -3e-4
FC25:n Ring detector at side
FZ25:n   2 4  0
dd25 -3e-4
FC55 Pinhole image projection radiograph at side
```

```
FIP55:n 0 50 2  0   0 0 2  5  0   40
c55   -3 99i 3
FS55  -3 99i 3
FC65 Transmitted image projection radiograph on axis
FIR65:n 0 0 40  0   0 0 0  0  0   0
c65   -20 99i 20
FS65  -20 99i 20
FC75 Transmitted image projection radiograph on side
FIR75:n 4 0 0   0   0 0 0  0  0   0
c75   -10 99i 30
FS75  -20 99i 20
c
FC104 MCNP5 style neutron FMESH tally
FMESH104:n geom=rec origin -5 -5 -30
        IMESH  5      IINTS 100
        JMESH  5      JINTS 100
        KMESH 30 31   KINTS 1 1
FC114 MCNP5 style proton FMESH tally
FMESH114:h geom=rec origin -5 -5 -30
        IMESH  5      IINTS 100
        JMESH  5      JINTS 100
        KMESH 30 31   KINTS 1 1
FC124 MCNP5 style neutron source FMESH tally
FMESH124:n geom=rec origin -5 -5 -60 type source
        IMESH  5      IINTS 100
        JMESH  5      JINTS 100
        KMESH 60      KINTS 1
c
TMESH
  RMESH201:n FLUX  TRAKS
    CORA201   -5 99I 5
    CORB201   -5 99I 5
    CORC201  -30 30 31
  RMESH211:h FLUX  TRAKS
    CORA211   -5 99I 5
    CORB211   -5 99I 5
    CORC211  -30 30 31
  RMESH232    N
    CORA232   -5 99I 5
    CORB232   -5    5
    CORC232  -60 30i -.1 60i 4 10i 20
ENDMD
MPLOT FREQ 1e4 plot pz 1 ex 5.5 la 0 1 tal201.1 col on
```

The following plot COMmand file shows how to plot the F15 point, FZ25 ring, F55, F65, F75 radiography tallies and FMESH104, FMESH114, and FMESH124 tallies. Plotting the TMESH tallies is problematic and is dealt with thereafter.

```
rmc j161.m
tal 15 la "F15 point" tfc m cop tal 25 la "FZ25 Ring"
pause
tal 55 free sc
pause
tal 65 free sc
pause
tal 75 free sc
pause
run j161.r
plot
fmesh 104
pause
fmesh 114
pause
fmesh 124
pause
pause
end
end
```

Figure 2.34 is a coplot of the F:15 point detector tally and FZ25 ring detector tally from Example 2.22.

```
FC15:n Point detector at side
F15:n   4 0 2   0
dd15  -3e-4
FC25:n Ring detector at side
FZ25:n    2 4   0
dd25  -3e-4
```

The point and ring detectors converge within statistical uncertainty because of the cylindrical symmetry of the problem. In such problems, the ring detector is more efficient and converges with a higher FOM. Note that the SDEF source requires the ARA entry to give the area of the plane-wave source and enable next-event estimators such as the point detector, ring detector, and radiography tallies.

The pinhole radiograph tally (FIP) is

```
FC55 Pinhole image projection radiograph at side
FIP55:n 0 50 2  0   0 0 2  5   0   40
c55    -3 99i 3
FS55   -3 99i 3
```

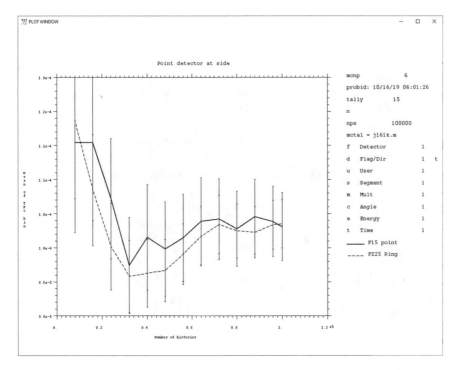

Fig. 2.34 Coplot of F15 point detector and equivalent FZ25 ring detector tallies. The plot command is "rmc j161k.m tal 15 la "F15 point" tfc m cop tal 25 la "FZ25 Ring""

The FIP55 tally is a pinhole radiography tally. The pinhole is like the focal point in a camera. The pinhole radiography tally captures the image of a particles emitted from or reflected by a region. The first three entries are the coordinates of the pinhole (0,50,2). The 4th entry is unused. The 5th–7th entries are the center of the object (0,0,2). The tally axis is the vector from the object (0,0,2) to the pinhole, (0,50,2)—namely the y-axis—normal to the incident beam axis. No collision or source points outside a 5 cm radial collimator (8th entry) contribute to the radiograph, saving computational time. The 9th entry is the radius of the pinhole (0=perfect pinhole). The image plane is 40 cm (10th entry) beyond the pinhole in the tally axis (y-axis) direction. The s- and t-axes are x and y lengths of the radiograph plane and are specified on the FS (segment) and C (cosine) cards as dimensions in cm. TALNP prevents excessive printout to the OUTP file. The plot is illustrated in Fig. 2.35.

Any tally can be plotted as a two-dimensional array of any two of the 8 tally dimensions. For radiography tallies, it is convenient to plot the cosine and segment dimension. The "free" tally plot command can be in one dimension as in "free s" or "free c," or in two dimensions as in "free sc."

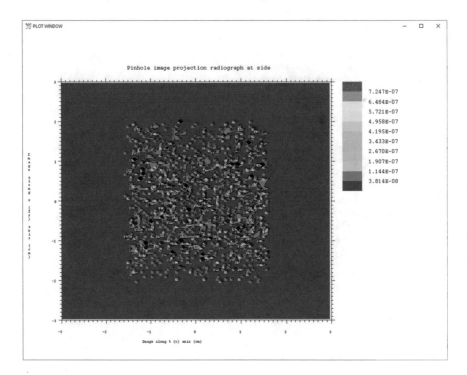

Fig. 2.35 Pinhole radiography tally F55 of Example 2.22. The image plane is a side (y-axis) view of NE213 detector secondary and scattered neutrons. The plot command is "tal 55 free sc"

Example 2.22 has two transmitted image (FIR) rectangular coordinate radiography tallies. Cylindrical transmitted image radiography tallies, FIC, are available but not illustrated.

```
FC65 Transmitted image projection radiograph on axis
FIR65:n 0 0 40  0  0 0 0  0  0  0
c65    -20 99i 20
FS65   -20 99i 20
FC75 Transmitted image projection radiograph on side
FIR75:n 4 0 0  0  0 0 0  0  0  0
c75    -10 99i 30
FS75   -20 99i 20
```

The first three entries are the coordinates of the center of the image plane, which is normal to the tally axis. The 4th entry is unused. The tally axis is the vector from the image center (5th–7th entries) (0,0,0) to the image plane center [(0,0,40) for FIR65 and (4,0,0) for FIR75]. The last 8th entry (0) tallies both direct and indirect (scattered) contributions; = 1 for scattered only. The 9th entry is the radial field of view (= 0 for infinite). The 10th entry scores to the center of each s- and t-axis grid point rather than to a random point in the grid.

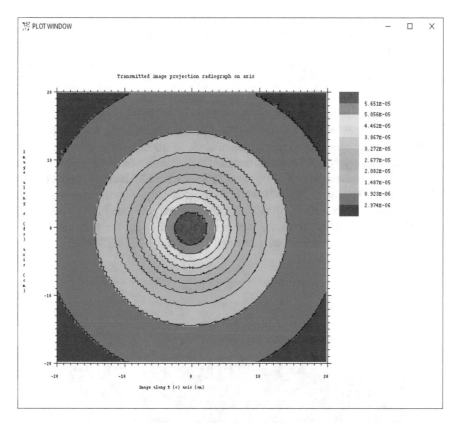

Fig. 2.36 Transmitted image radiography tally F65 of Example 2.22. The image plane is in the SDEF source direction (on z-axis) and behind the object. The tally plot command is "tal 65 free sc"

The transmitted image radiography tallies of Example 2.22 are plotted below: Fig. 2.36 is the image from the F65 on-axis tally; Fig. 2.37 is the image from the F75 side-view tally.

The MCNP5 style mesh tallies in Example 2.22 show the neutron flux (FMESH104), proton flux (FMESH114), and neutron source (FMESH124) integrated over the z-axis source direction and in the x–y plane:

```
FC104 MCNP5 style neutron FMESH tally
FMESH104:n geom=rec origin -5 -5 -30
        IMESH  5      IINTS 100
        JMESH  5      JINTS 100
        KMESH 30 31   KINTS 1 1
FC114 MCNP5 style proton FMESH tally
FMESH114:h geom=rec origin -5 -5 -30
        IMESH  5      IINTS 100
        JMESH  5      JINTS 100
        KMESH 30 31   KINTS 1 1
```

```
FC124 MCNP5 style neutron source FMESH tally
FMESH124:n geom=rec origin -5 -5 -60 type source
          IMESH  5       IINTS 100
          JMESH  5       JINTS 100
          KMESH 60       KINTS 1
```

These tallies can be plotted from the MCNP RUNTPE continuation file in the geometry plot mode simply by clicking the FMESH button in the lower left of the MCNP geometry plot or by entering commands at the command prompt or in a COMmand file. The MCNP code automatically determines the view as through the middle of the largest mesh span. Figure 2.38 shows the FMESH104 neutron mesh tally. Figure 2.39 shows FMESH114 proton mesh tally. Figure 2.40 shows the FMESH124 neutron source mesh tally. In each of these plots, the PLOT> command "VIEWPORT SQUARE" from the terminal was used to get the plot without the interactive geometry plot buttons showing. The command "VIEWPORT RECT"

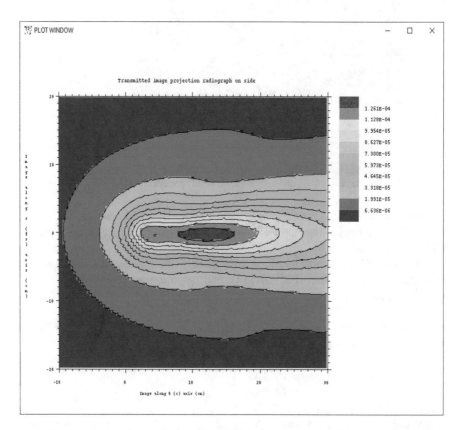

Fig. 2.37 Transmitted image radiography tally F75 of Example 2.22. The image plane is a side view normal to the x-axis, normal to the source direction z axis direction, and shows a side view of neutron emission from the object. The tally plot command is "tal 75 free sc"

restores the rectangular plot plane and "INTERACT" switches from PLOT> command mode back to interactive mode.

The MCNPX style TMESH tallies in Example 2.22 enable more plotting capabilities than the FMESH tallies. Tracks, populations, doses, DXTRAN, point detector, multiple particle energy deposition, and more can be plotted in addition to fluxes and source points. Unfortunately, the interface is inconsistent with the FMESH interface and harder to use. The TMESH tallies start with "TMESH" and end with "ENDMD" and must be all together:

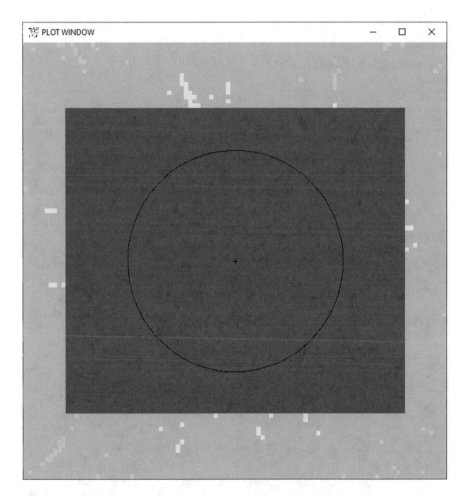

Fig. 2.38 FMESH neutron flux tally F104, which excludes the source beam and consists of secondary neutrons. The plot is made by clicking the FMESH button on the lower left of the geometry plot or by entering the command "FMESH 104"

```
TMESH
  RMESH201:n FLUX   TRAKS
     CORA201    -5 99I 5
     CORB201    -5 99I 5
     CORC201   -30  30  31
  RMESH211:h FLUX   TRAKS
     CORA211    -5 99I 5
     CORB211    -5 99I 5
     CORC211   -30  30  31
  RMESH232     N
     CORA232    -5 99I 5
     CORB232    -5     5
     CORC232   -60 30i -.1 60i  4  10i  20
ENDMD
```

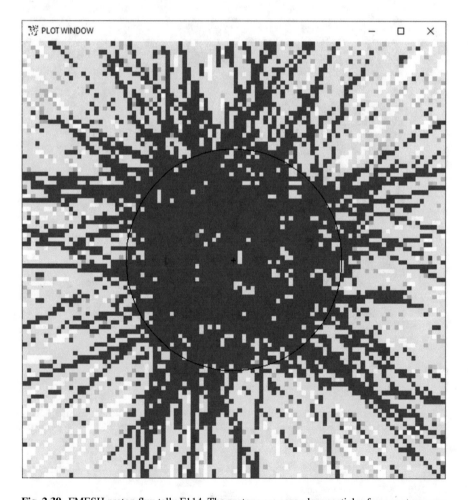

Fig. 2.39 FMESH proton flux tally F114. The protons are secondary particles from neutron reactions. The plot is made by clicking the FMESH button on the lower left of the geometry plot or by entering the command "FMESH 114"

There are also bugs that make entering PLOT> commands ineffective. Entering commands from a COMmand file, from the terminal, or in the lower left of the interactive geometry plot where it says "click here or picture or menu" does not work unless the lower right button "TAL" in the interactive geometry plot reading from the RUNTPE is first toggled. After that, the "TAL," "PAR," and "N" buttons can be used or the TMESH plot commands may be entered. Note that the FMESH button must be clicked off. (There are TMESH plotting examples in the MCNP manual [2] Sect. 6.4.4.)

The neutron flux and particle track rectangular RMESH201 mesh tallies are shown in the TMESH plots of Figs. 2.41 and 2.42. The X–Y view is normal to the source z-axis. The proton flux and particle track rectangular RMESH211 mesh tallies are shown in the TMESH plots of Figs. 2.43 and 2.44. Again, the X–Y view is normal to the source z-axis. The neutron source rectangular RMESH232 mesh tally is shown

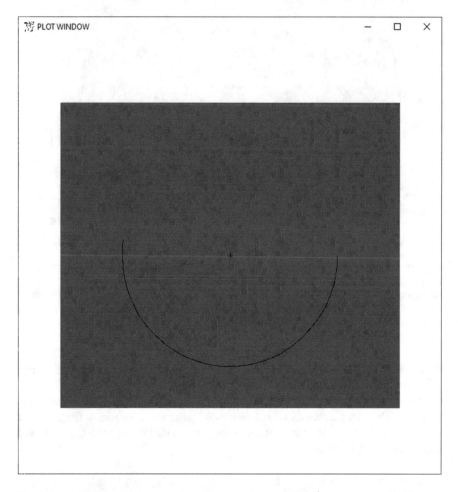

Fig. 2.40 FMESH neutron source tally F124. The neutrons are from proton collisions. The $z = 0$ plot plane is coincident with the bottom of the NE213 detector object, resulting in the semi-circle surface boundary. The plot is made by clicking the FMESH button on the lower left of the geometry plot or by entering the command "FMESH 124"

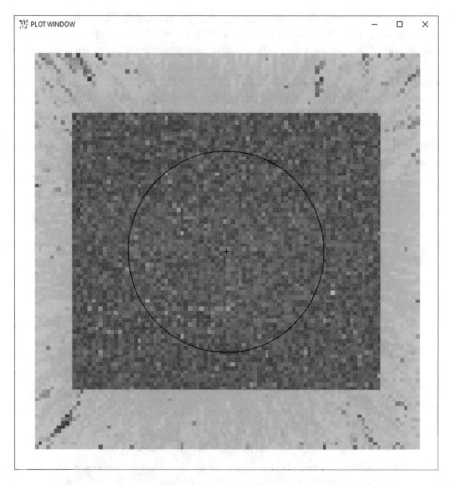

Fig. 2.41 TMESH plot of rectangular mesh tally RMESH201 neutron flux made from RUNTPE in the geometry plot mode. The view is down the source z-axis, "pz 1 ex 5.5 or 0 0 1." The plot command after clicking off FMESH and cell labels and entering VIEWPORT SQUARE is "la 0 1 tal201.1 la 0 0." The 0.1 gets the first PAR entry, namely flux. The first LA command gets the tally for coloring and the second turns off cell labels

in the TMESH plot of Fig. 2.45. The X–Z view is normal to y-axis with the source z-axis in the vertical direction.

The run-time plotting for TMESH tallies utilizes the MPLOT command in the input file. The plot picture will flash on the computer terminal with a frequency of every FREQ = 1.E–4 histories. The remaining commands specify the plot view and which TMESH tally to plot.

```
MPLOT FREQ 1e4 plot pz 1 ex 5.5 la 0 1 tal201.1 col on
```

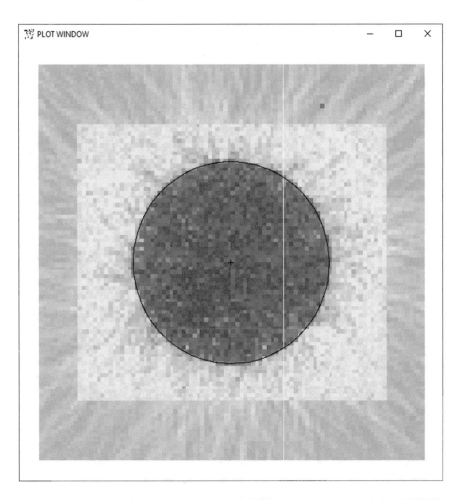

Fig. 2.42 TMESH plot of rectangular mesh tally RMESH201 neutron tracks made from RUNTPE in the geometry plot mode. The plot view is down the source z-axis with geometry plot command "pz 1 ex 5.5 or 0 0 1." The plot command after clicking off FMESH and cell labels and entering "VIEWPORT SQUARE" is "la 0 1 tal201.2 la 0 0." The 0.2 gets the second PAR entry, namely TRAKS. The first LA command gets the tally for coloring and the second turns off cell labels

2.6 Statistics and Convergence

Example 2.23 is a thin, 2 cm-thick disc with a huge 1×10^{10} cm radius.

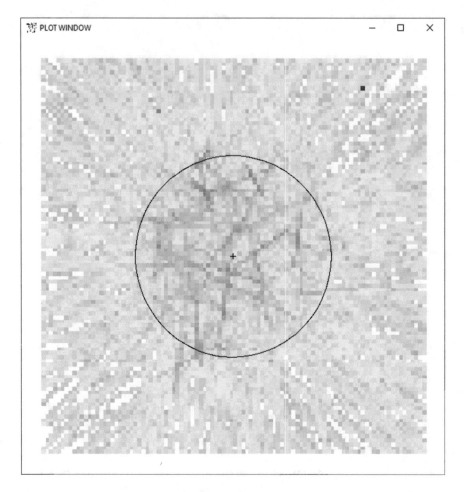

Fig. 2.43 TMESH plot of rectangular mesh tally RMESH211 proton flux made from RUNTPE in the geometry plot mode. The plot view is "pz 1 ex 5.5 or 0 0 1," namely down the source z-axis. In the interactive mode, FMESH must be off, cell labels L2 off, and the lower right "TAL" and "N" buttons must be cycled through until tally 211 is the edit quantity (middle left). Then click PLOT> and enter "VIEWPORT SQUARE" and "la 0 1 tal211.1 la 0 0." The 0.1 gets the first PAR entry, namely flux. The first LA command gets the tally for coloring and the second turns off cell labels

Example 2.23 Demonstration of False Convergence and Ten Statistical Tests

```
Isotropic source in RCC disc
1 0 -1 imp:n=1
2 0  1 imp:n=0

1 RCC 0 0 -1  0 0 2  1e10
```

```
nps 2e4
prdmp 2j 1
print
c
fc11 Neutrons crossing surface
f11:n (1.1 1.2 1.3) 1.1 1.2 1.3
ft11 frv 0 0 1
c11 -.95 38i 1 T
fq11 C F
c
fc4 Neutron flux in RCC disc
f4:n 1
sd4 1
c
SDEF vec 0 0 1 dir=d1
si1 -1     9ilog        -1e-10 1e-10 9ilog 1
sp1  0 .9 8ilog 9e-10 2e-10 9e-10 8ilog .9
```

The problem is contrived to demonstrate the ten MCNP statistical tests that can help identify false convergence. Tally F4 of this problem passes nine of the ten MCNP statistical tests although it is falsely converged after 20000 histories to an estimate of $9.19 \pm 5.9\%$. The correct answer is 24.06 neutrons, more than a factor of two different. Sometimes false convergence gives estimates that are orders of magnitude wrong.

False convergence is caused by undersampling important regions of problem space. Undersampling can occur if an important region of the geometry is not sufficiently sampled. Undersampling also occurs if important physics, such as neutron resonances or photon coherent scatter, are improperly sampled. In Example 2.23, undersampling is caused by insufficiently sampling source angle.

The source in Example 2.23 is an isotropic source because the number of neutrons started in each cosine bin from $-1 < \mu < 1$ is proportional to the size of the cosine bin. This is equivalent to these other forms of an isotropic source:

```
SDEF
```

or

```
SDEF vec 0 0 1 dir=d1
si1 -1  1
sp1  0  1
```

Figure 2.46 shows that the source is indeed isotropic. It is a plot of the tally F11 neutron current in cosine bins.

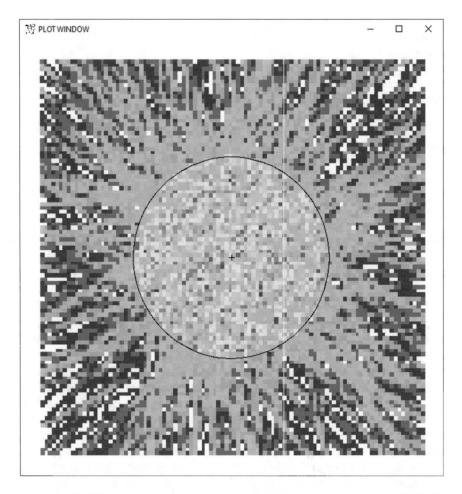

Fig. 2.44 TMESH plot of rectangular mesh tally RMESH211 proton tracks made from RUNTPE in the geometry plot mode.The plot view is "pz 1 ex 5.5 or 0 0 1," namely down the source z-axis. In the interactive mode, FMESH must be off, cell labels L2 off, and the lower right "TAL" and "N" buttons must be cycled through until tally 211 is the edit quantity (middle left). Then click PLOT> and enter "VIEWPORT SQUARE" and "la 0 1 tal211.2 la 0 0." The 0.2 gets the first PAR entry, namely flux. The first LA command gets the tally for coloring and the second turns off cell labels

The tally plot command is

```
rmc j20k.m tal 11 free c la "20k" linlin cop rmc j8.m tal 11
la "1e8"
```

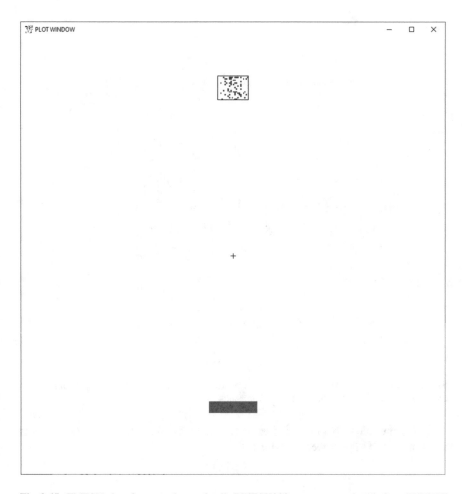

Fig. 2.45 TMESH plot of rectangular mesh tally RMESH232 neutron source made from RUNTPE in the geometry plot mode. The plot view is "py 0 ex 34 or 0 0 -25." Click off FMESH and cell labels. Click on TAL and N until tal232.1 becomes the edit quantity. Click PLOT> and enter "VIEWPORT SQUARE" to get the picture without buttons. The SDEF source is seen in the –Z mesh region at the bottom of the plot. The source plane is a line in this view, but appears like a rectangle with a thickness of the tally mesh. The red color is the maximum source strength. Secondary neutrons from protons are seen at the NE213 detector at the top of the plot. These secondary neutrons appear as blue dots because of their lower intensity

The MCTAL file j20k.m was created with NPS 2e4 and the MCTAL file j8.m was created with NPS 1e8. The number of neutrons in each of the 40 equal cosine bins is the same within statistical uncertainty, which is an isotropic distribution.

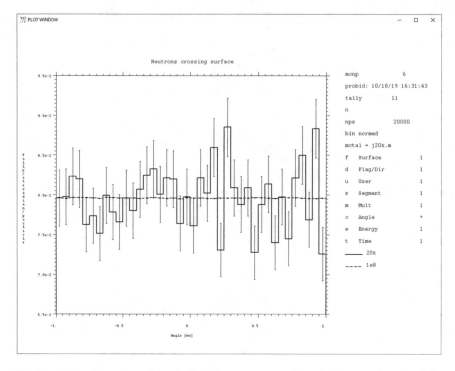

Fig. 2.46 Tally plot of Example 2.23 tally F11 neutron current in cosine bins; number of particles crossing all surfaces of the disc relative to the z-axis

The output file tally fluctuation chart for the F4 flux tally shows the tally results as a function of the number of histories:

	tally			4	
nps	mean	error	vov	slope	fom
2000	8.2767E+00	0.1805	0.5439	1.9	117916
4000	7.8166E+00	0.1082	0.3385	2.0	163975
6000	8.4977E+00	0.0983	0.1613	2.0	132371
8000	8.7945E+00	0.0949	0.1873	2.3	106554
10000	8.8567E+00	0.0820	0.1405	2.2	114229
12000	8.8204E+00	0.0759	0.1160	2.2	133236
14000	8.7715E+00	0.0679	0.1014	2.5	138769
16000	8.7582E+00	0.0617	0.0886	2.4	143993
18000	9.0119E+00	0.0592	0.0715	2.3	136872
20000	9.1865E+00	0.0589	0.0726	2.3	122876

The "mean" is the flux estimate of the answer. The "error" is the relative error, which is the standard deviation of the mean divided by the mean. The "vov" is the variance of the variance, or the relative error of the relative error. The "slope" is the rate of convergence. The "fom" is the Figure of Merit.

The tally printed in the output file is

```
1tally         4         nps =       20000              Neutron flux in RCC disc
       tally type 4    track length estimate of particle flux. units    1/cm**


   cell   1
                       9.18646E+00 0.0589


results of 10 statistical checks for the estimated answer for the tally
fluctuation chart (tfc) bin of tally         4


    tfc bin      --mean--        ---------relative error---------
    behavior     behavior        value   decrease   decrease rate


    desired      random          <0.10     yes      1/sqrt(nps)
    observed     random           0.06     yes          yes
    passed?       yes             yes       yes          yes

    ==============================================================


    ----variance of the variance----      --figure of merit--     -pdf-
    value   decrease   decrease rate        value    behavior     slope
    <0.10     yes         1/nps             constant   random      >3.00
     0.07     yes          yes             constant   random       2.32
     yes      yes          yes                yes        yes         no


number of nonzero history tallies = 20000
efficiency for the nonzero tallies  = 1.0000
history number of largest  tally  = 18954
largest  unnormalized history tally = 4.42936E+03
(largest  tally)/(average tally)   = 4.82162E+02
(largest  tally)/(avg nonzero tally)= 4.82162E+02
```

As shown in the tally fluctuation chart at the end of the output file (above), the mean value, 9.18646E+00, is constant within statistics—or "random"—passing the first of the ten statistical tests. The relative error, $E = 0.0589$, is <10%, decreasing, and sufficiently decreasing as 1/sqrt(NPS), thus passing the 2nd–4th statistical tests. For next-event estimator tallies, F5 point detectors, the passing criteria is a relative error <5%. The variance of the variance, VOV = 0.0726, is less than 10%, decreasing, and sufficiently decreasing as 1/NPS, passing the 5th–7th statistical tests. For convergence, the VOV test is more important and more sensitive than the relative error test. The Figure of Merit, FOM = 122876, is $1/(TE^2)$, where T is the computation time and E is the relative error. Whereas E^2 is proportional to 1/NPS and T is proportional to NPS, the FOM should be a constant. Note that the FOM will vary from one computer to another because T is machine dependent, so FOM can be compared only on the same computer, MCNP version, and installation. Because the FOM is sufficiently constant, the 8th and 9th convergence tests are passed. Only the 10th test—the slope test—fails, as indicated by "no."

The slope test estimates the rate of large scores that contribute to a tally. If a region of phase space is undersampled, then its score to the tally will be large when it finally is sampled because the Monte Carlo calculation is unbiased and the frequency of large scores times the magnitude of those scores should be the expectation value—or mean score—of the problem.

Figure 2.47 in the output shows how the slope is calculated.

All scores from 1e−30 to 1e30 are put in 600 logarithmic bins according to score size. Then the first nonzero score bin, in this case 1.26+00, to the highest, in this case 5.01+03, is printed vertically with the number of scores in each bin. The "num den" is the number density, or the number of scores, divided by the size of the bin. Then the logarithm of this density is taken and "*" (asterisks) are printed to represent the "log den" magnitude. The slope is the curve fitted to the 200 highest scores, indicated by the "s." Print Table 161 shows 6 scores >2.00+03, which flatten the slope to its value of 2.32 and are 200 times the average score of 9.18. The largest unnormalized history tally is $4.42936E+03$ (from print table 160), which is $4.82162E+02$ times larger than the average tally.

To pass all convergence tests, the slope should be >3 [11]. To increase the slope, the parts of the problem that cause the large scores need to be sampled more often. To determine the cause of the largest score, one can rerun the largest history (the relevant history number is found in print table 160):

```
NPS 1
RAND HIST=18954
DBCN   2J  1  1
```

NPS runs a single history. The random number sequence of history 18954 is specified via the RAND card. An event log for the 1st through the 1st history is specified by the DBCN card. Thus, History 18954 is repeated, and from the 1st 50-history Print Table 110 and the event log, the cause of the large score becomes apparent. The cosine of the source in the z-direction, W, is very small and leads to a track length of $\lambda = 4.42936E+03$ cm before the particle escapes. The F4 flux tally is F4 = λ/V and $V = 1$ because of the SD card. In fact, a track length comparable to the disc radius, 1e8, is possible. Thus z-axis cosine bins near zero need to be sampled better. Setting a source bias of

```
    sb1  0  1  20r
```

causes the SDEF source to start more particles in the small z-angle directions, as seen in Fig. 2.48 from Print Table 10 after the SB bias is added. This source direction bias enables the problem to pass all statistical checks and converge to the correct answer very quickly. (Note the weight multiplier keeps this a fair physical representation of an isotropic source.)

```
1unnormed tally density for tally 4          nonzero tally mean(m) = 9.186E+00    nps = 20000              print table 161

abscissa              ordinate   log plot of tally probability density function in tally fluctuation chart bin(d=decade,slope= 2.3)
 tally   number num den log den:d------------d-------------d-------------d-------------d-------------d-------------d-------------d-
1.26+00     4182 8.08-01  -0.093 *************|*************|*************|*************|*************|*************|*************|*
1.58+00     3178 4.87-01  -0.312 *************|*************|*************|*************|*************|*************|*************|
2.00+00     2544 3.10-01  -0.509 *************|*************|*************|*************|*************|*************|***********  |
2.51+00     2015 1.95-01  -0.710 *************|*************|*************|*************|*************|*************|********     |
3.16+00     1650 1.27-01  -0.897 *************|*************|*************|*************|*************|*************|******       |
3.98+00     1391 8.49-02  -1.071 *************|*************|*************|*************|*************|*************|***          |
5.01+00     1004 4.87-02  -1.312 *************|*************|*************|*************|*************|*************|             |
6.31+00      885 3.41-02  -1.467 *************|*************|*************|*************|*************|***********  |             |
7.94+00      662 2.03-02  -1.693 *************|*************|*************|*************|*************|********     |             |
1.00+01      514 1.25-02  -1.903 mmmmmmmmmmmmm|mmmmmmmmmmmmm|mmmmmmmmmmmmm|mmmmmmmmmmmmm|mmmmmmmmmmmmm|mmmmm        |             |
1.26+01      410 7.92-03  -2.101 *************|*************|*************|*************|*************|**           |             |
1.58+01      350 5.37-03  -2.270 *************|*************|*************|*************|*************|             |             |
2.00+01      234 2.85-03  -2.545 *************|*************|*************|*************|**********   |             |             |
2.51+01      192 1.86-03  -2.731 *************|*************|*************|*************|*******      |             |             |
3.16+01      133 1.02-03  -2.990 *************|*************|*************|*************|****         |             |             |
3.98+01      127 7.76-04  -3.110 *************|*************|*************|*************|**           |             |             |
5.01+01       92 4.46-04  -3.350 *************|*************|*************|*************|             |             |             |
6.31+01       91 3.51-04  -3.455 *************|*************|*************|***********  |             |             |             |
7.94+01       63 1.93-04  -3.715 *************|*************|*************|********     |             |             |             |
1.00+02       51 1.24-04  -3.907 *************|*************|*************|*****        |             |             |             |
1.26+02       53 1.02-04  -3.990 *************|*************|*************|****        s|             |             |             |
1.58+02       38 5.83-05  -4.234 *************|*************|*************|           s |             |             |             |
2.00+02       30 3.66-05  -4.437 *************|*************|************ |          s  |             |             |             |
2.51+02       21 2.03-05  -4.692 *************|*************|********     |        s    |             |             |             |
3.16+02       15 1.15-05  -4.938 *************|*************|*****        |      s      |             |             |             |
3.98+02       16 9.77-06  -5.010 *************|*************|****         |    s        |             |             |             |
5.01+02       18 8.73-06  -5.059 *************|*************|***          |  s          |             |             |             |
6.31+02        8 3.08-06  -5.511 *************|***********  |             s             |             |             |             |
7.94+02        8 2.45-06  -5.611 *************|*********    |           s |             |             |             |             |
1.00+03        6 1.46-06  -5.836 *************|******       |        s    |             |             |             |             |
1.26+03        6 1.16-06  -5.936 *************|*****        |      s      |             |             |             |             |
1.58+03        3 4.60-07  -6.337 *************|             |   s         |             |             |             |             |
2.00+03        4 4.87-07  -6.312 *************|             s             |             |             |             |             |
2.51+03        0 0.00+00   0.000              |           s |             |             |             |             |             |
3.16+03        3 2.31-07  -6.637 *********    |        s    |             |             |             |             |             |
3.98+03        1 6.11-08  -7.214 *            |     s       |             |             |             |             |             |
5.01+03        2 9.70-08  -7.013 ****         | s           |             |             |             |             |             |
 total     20000 1.00+00         d------------d-------------d-------------d-------------d-------------d-------------d-------------d-
```

Fig. 2.47 Print Table 161 is the unnormed tally density function plot. The abscissa is the number of unnormalized nonzero scores. Following columns are the number of scores, the number density (the number of scores divided by the width of tally bins in column 1), the log of that number, and then asterisks proportionate to the number of scores. The slope fit is denoted by "s"

```
probability distribution    1 for source variable dir    print table 10
biased histogram distribution
```

source entry	source value	cumulative probability	biased cumulative	probability of bin	biased probability	weight multiplier
1	-1.00000E+00	0.000000E+00	0.000000E+00	0.000000E+00	0.000000E+00	1.000000E+00
2	-1.00000E-01	4.500000E-01	4.761905E-02	4.500000E-01	4.761905E-02	9.450000E+00
3	-1.00000E-02	4.950000E-01	9.523810E-02	4.500000E-02	4.761905E-02	9.450000E-01
4	-1.00000E-03	4.995000E-01	1.428571E-01	4.500000E-03	4.761905E-02	9.450000E-02
5	-1.00000E-04	4.999500E-01	1.904762E-01	4.500000E-04	4.761905E-02	9.450000E-03
6	-1.00000E-05	4.999950E-01	2.380952E-01	4.500000E-05	4.761905E-02	9.450000E-04
7	-1.00000E-06	4.999995E-01	2.857143E-01	4.500000E-06	4.761905E-02	9.450000E-05
8	-1.00000E-07	5.000000E-01	3.333333E-01	4.500000E-07	4.761905E-02	9.450000E-06
9	-1.00000E-08	5.000000E-01	3.809524E-01	4.500000E-08	4.761905E-02	9.450000E-07
10	-1.00000E-09	5.000000E-01	4.285714E-01	4.500000E-09	4.761905E-02	9.450000E-08
11	-1.00000E-10	5.000000E-01	4.761905E-01	4.500000E-10	4.761905E-02	9.450000E-09
12	1.00000E-10	5.000000E-01	5.238095E-01	1.000000E-10	4.761905E-02	2.100000E-09
13	1.00000E-09	5.000000E-01	5.714286E-01	4.500000E-10	4.761905E-02	9.450000E-09
14	1.00000E-08	5.000000E-01	6.190476E-01	4.500000E-09	4.761905E-02	9.450000E-08
15	1.00000E-07	5.000001E-01	6.666667E-01	4.500000E-08	4.761905E-02	9.450000E-07
16	1.00000E-06	5.000005E-01	7.142857E-01	4.500000E-07	4.761905E-02	9.450000E-06
17	1.00000E-05	5.000050E-01	7.619048E-01	4.500000E-06	4.761905E-02	9.450000E-05
18	1.00000E-04	5.000500E-01	8.095238E-01	4.500000E-05	4.761905E-02	9.450000E-04
19	1.00000E-03	5.005000E-01	8.571429E-01	4.500000E-04	4.761905E-02	9.450000E-03
20	1.00000E-02	5.050000E-01	9.047619E-01	4.500000E-03	4.761905E-02	9.450000E-02
21	1.00000E-01	5.500000E-01	9.523810E-01	4.500000E-02	4.761905E-02	9.450000E-01
22	1.00000E+00	1.000000E+00	1.000000E+00	4.500000E-01	4.761905E-02	9.450000E+00

Fig. 2.48 The source distribution is shown in optional (requires PRINT card) Print Table 10 of the output file. Each source bin is sampled with the same probability, but because the bin widths are different, the weights vary by nine orders of magnitude. Consequently, the central, very narrow bin—normal to the source direction axis—is sampled nine orders of magnitude more to sufficiently achieve convergence

Thus, the ten statistical tests identify false or poor convergence. The solution to false or poor convergence is variance reduction, which results in sampling the more important parts of the problem more frequently at the expense of less important parts.

What does one do when some tests fail? The ten statistical tests are themselves estimated quantities and can sometimes be misleading:

- If the first test of the mean having a constant value fails, check the tally fluctuation charts at the end of the output file to see if it is really constant. In rare cases, the MCNP fitting routine fails.
- If the 2nd test of the relative error >10% (5% for point detectors/next-event estimators) fails, run the problem for more particles or computer time. Note that if the slope <3, the relative error does not exist, and running longer may not help.
- If the 3rd and 4th tests of the relative error behaving as 1 / SQRT (NPS) fail, in rare cases, the MCNP code is unable to fit the curve properly.
- If the 5th–7th tests of the variance of the variance, VOV, fail, be aware that VOV does not exist if the slope is <3 and can be ignored in those cases. Otherwise, run longer, or check the tally fluctuation chart for a linearly decreasing VOV.
- If the 8th and 9th tests of the FOM fail, check the tally fluctuation chart to see if it really is not constant. Note that—in the early stages of a problem—the MCNP code does extra checking for the first 50 to a few thousand histories. Thus, on short runs of merely thousands of histories, the FOM will increase as the checks finish. Also, the FOM is based on computer time so that very short problems of a few seconds often have difficulty passing the FOM tests.
- If the 10th slope test fails, then the important parts of the problem have probably not been sampled well enough. Note that the slope test requires at least 600 histories and the ability to fit the slope to the 200 largest scores, so a slope of zero is observed for short problems. These short problems need to be run longer for more histories. The user should be aware of what the largest possible score to any tally could be and where large scores are likely. Then, variance reduction should be used to sample whatever causes the largest score and other likely large scores. Note that improper use of variance-reduction methods can also lead to failing the slope test, which often indicates poor convergence right from the start of a problem before the other tests are passed. The slope test is the most challenging test to understand and use to improve convergence.

References

1. MCNP6 website: https://mcnp.lanl.gov/index.shtml
2. C.J. Werner (ed.), *MCNP User's Manual Code Version 6.2*, LA-UR-17-29981 (October 27, 2017)
3. C.J. Werner, J.S. Bull, C.J. Solomon et al., *MCNP Release Notes Code Version 6.2*, LA-UR-18-20808 (2018)

4. Nuclear Data Team, XCP-5, *Listing of Available ACE Data Tables*, LA-UR-17-20709 (October 12, 2017)
5. F.H. Frohner, Evaluation of ^{252}Cf Prompt Fission Neutron Data from 0 to 20 MeV by Watt Spectrum Fit. Nucl. Sci. Eng. **106**, 345–352 (1992)
6. John R. Lamarsh, *Introduction to Nuclear Reactor Theory*, Addison-Wesley, Reading Massachusetts (1966)
7. J.S. Hendricks, G.W. McKinney, *Pulse-Height Tallies with Variance Reduction*, Los Alamos National Laboratory Report LA-UR-04-8431, Monte Carlo 2005, Chattanooga, TN (April 17–21, 2005)
8. G.F. Knoll, *Radiation Detection and Measurement* (Wiley, Hoboken, NJ, 2000)
9. T. Wilcox, *"MCNP Advanced Tallies Tutorial" Topical Meeting of the Radiation Protection and Shielding Division of the American Nuclear Society*, Topical Meeting September 2014, Knoxville, TN. Los Alamos National Laboratory Report LA-UR-14-27128 (2014)
10. J.L. Conlin, F.B. Brown, A.C. Kahler III, M.B. Lee, D.K. Parsons, M.C. White, *Updating the Format of ACE Data Tables*, Los Alamos National Laboratory Report LA-UR-12-22033 (2012)
11. *MCNP5 Theory Manual Chapter 2, Section VI, Subsection H.7*, pp. 2–127. https://laws.lanl.gov/vhosts/mcnp.lanl.gov/pdf_files/la-ur-03-1987.pdf

Chapter 3
Examples for Nuclear Safeguards Applications

3.1 Fuel Assembly in Water Tank

3.1.1 Problem Description

The MCNP input file of Example 3.1 describes a 13 × 13 fuel assembly in a cylindrical water tank. It shows the MCNP repeated structures lattice capability. The aim of the input file is to calculate the induced fission in the enriched and depleted uranium pins in the fuel assembly in the water tank when the starting source of neutrons is spontaneous fission.

Example 3.1 Fuel Assembly in Water Tank

```
Fuel Assembly in Water Tank
c *************************** cells ***************************
11 313 -10.44   -71     u=1 imp:n=1 $ Fuel
12 201   -6.55   -72 71   u=1 imp:n=1 $ Cladding
13 101   -1.00        72   u=1 imp:n=1 $ Outside u=1 world
21 101   -1.00       -73  u=5 lat=1          imp:n=1 $ Infinite lattice
         fill=-6:6 -6:6 0:0
          5 1 1 1 1 1 1 1 1 8 8 8 8
          1 1 1 1 1 1 1 1 1 1 1 1 1
          1 1 1 1 1 1 1 1 1 1 1 1 1
          1 1 1 1 1 1 1 1 1 1 1 1 1
          1 1 1 1 1 1 1 1 1 1 1 1 1
          1 1 1 1 1 1 1 1 1 1 1 1 1
          1 1 1 1 1 1 1 1 1 1 1 1 1
```

Supplementary Information The online version contains supplementary material available at [https://doi.org/10.1007/978-3-031-04129-7_3].

J. S. Hendricks et al., *Monte Carlo N-Particle Simulations for Nuclear Detection and Safeguards*, https://doi.org/10.1007/978-3-031-04129-7_3

```
           1 1 1 1 1 1 1 1 1 1 1 1
           1 1 1 1 1 1 1 1 1 1 1 1
           1 1 1 1 1 1 1 1 1 1 1 1
           1 1 1 1 1 1 1 1 1 1 1 1
           1 1 1 1 1 1 1 1 1 1 1 1
           1 1 1 1 1 1 1 1 1 1 1 1
31 0               -74          fill=5 imp:n=1 $ Assembly containing cell
32 101  -1.00    74 -75             imp:n=1 $ Water tank
99 0                   75           imp:n=0 $ Outside world
16 102  -10.5   -76             u=8 imp:n=1 $ Depleted uranium pin
17 101  -1.00          76       u=8 imp:n=1 $ Outside u=8 world

c ************************* surfaces *************************
71 RCC 0 0 -150    0 0 300   .4840          $ fuel cylinder
72 RCC 0 0 -150    0 0 300   .5590          $ clad cylinder
73 RPP -.75 .75   -.75 .75   -150 150       $ Fuel assembly box
74 RPP -9.75 9.75  -9.75 9.75  -150 150 $ Fuel assembly boundary
75 RCC 0 0 -200    0 0 400    30            $ water tank
76 RCC 0 0 -150    0 0 300   .7000          $ depleted uranium tube

c *************************** data ***************************
SDEF PAR=SF  pos=0 0 0  ext=d61  axs=0 0 1  rad=fcel=d50    cel=d63
si61 -150 150
sp61    0    1
si63 L    (11<21[-5:2   -6:-6 0:0]<31)
          (16<21[ 3:6   -6:-6 0:0]<31)
          (11<21[-6:6   -5:6  0:0]<31)
sp63      2.68 7R  5.80 3R  2.68 155R
DS50 S      51 7R     52 3R    51 155R
si51     0 .4840
sp51 -21 1
si52     0 .700
sp52 -21 1
print
nps 1e4
prdmp 2J 1
m102   92235.80c .003 92238.80c .997 8016.80c 2 $ depleted uranium
m201   40090 -0.505239   40091 -0.110180    40092 -0.168413
       40094 -0.170672   40096 -0.027496    nlib=80c    $ zirconium
m313   92235.80c -0.02759 92238.80c -0.85391 8016.80c -0.1185  $ UO2
m101   1001.80c 2 8016.80c 1                              $ water
mt101 lwtr
c
FC4 Induced fission neutrons in enriched and depleted fuel pins
F4:n   (11<(21[-5:2   -6:-6 0:0])<31) (11<(21[-6:6   -5:6  0:0])<31)
       (16<(21[ 3:6   -6:-6 0:0])<31)
```

```
FQ4  F  M
FM4  (-1  102  -6  -7)  (-1  313  -6  -7)
sd4  1  2R
```

Example 3.1 demonstrates how to use a lattice geometry to create a fuel assembly and how to calculate reaction rates (fission and fission-neutron production) in particular parts of the lattice. The different parts of the input file will be described, and then the output file and tally results will be explained.

The MCNP input file is divided into three sections. The cells are described, starting on the second line after the required title line down to the first blank line. The surfaces are then described down to the next blank line. All other information follows in the data section. Comment lines start with "c"; comments may also follow "$" on each line. Lines are limited to 128 characters wide and may be continued by either "&" or five spaces starting the next line.

The full geometry axial view is shown in Fig. 3.1. Figure 3.2 zooms into the lattice and shows the lattice indexing.

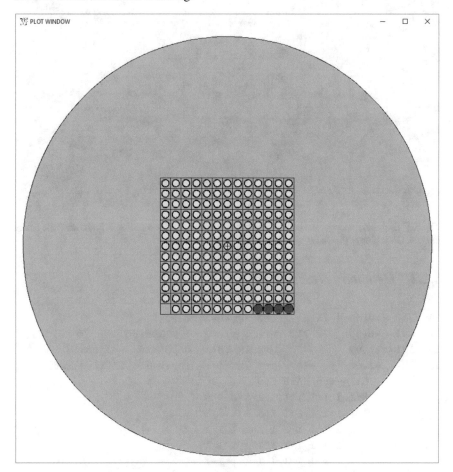

Fig. 3.1 MCNP plot of fuel assembly in water tank. The cells are colored by material: green is water (m101), yellow is LEU oxide (m313), and magenta is depleted uranium oxide (m102)

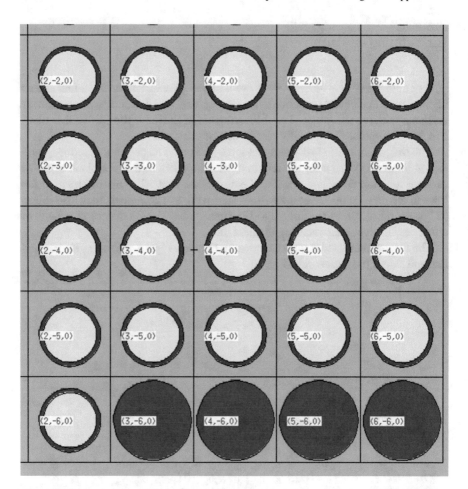

Fig. 3.2 Expanded view of the bottom right corner of the assembly in Fig. 3.1, showing the lattice indices. Dark blue is the cladding material (m201)

3.1.2 Geometry Description

```
11   313 -10.44   -71         u=1 imp:n=1  $ Fuel
12   201  -6.55   -72 71      u=1 imp:n=1  $ Cladding
13   101  -1.00        72     u=1 imp:n=1  $ Outside u=1 world
21   101  -1.00       -73     u=5 lat=1    imp:n=1 $ Infinite lattice
            fill=-6:6 -6:6 0:0
              5 1 1 1 1 1 1 1 1 8 8 8 8
              1 1 1 1 1 1 1 1 1 1 1 1 1
              1 1 1 1 1 1 1 1 1 1 1 1 1
              1 1 1 1 1 1 1 1 1 1 1 1 1
              1 1 1 1 1 1 1 1 1 1 1 1 1
```

```
              1 1 1 1 1 1 1 1 1 1 1 1
              1 1 1 1 1 1 1 1 1 1 1 1
              1 1 1 1 1 1 1 1 1 1 1 1
              1 1 1 1 1 1 1 1 1 1 1 1
              1 1 1 1 1 1 1 1 1 1 1 1
              1 1 1 1 1 1 1 1 1 1 1 1
              1 1 1 1 1 1 1 1 1 1 1 1
              1 1 1 1 1 1 1 1 1 1 1 1
31  0              -74         fill=5 imp:n=1 $ Assembly containing cell
32 101  -1.0    74 -75               imp:n=1 $ water tank
99  0                    75          imp:n=0 $ outside world
16 102  -10.5 -76               u=8  imp:n=1 $ Depleted uranium pin
17 101  -1.00        76         u=8  imp:n=1 $ Outside u=8 world
```

In this geometry, cell 11 is the fuel inside surface 71, cell 12 is the clad between surfaces 71 and 72, and cell 13 is outside surface 72 and extends out to infinity. These cells are all in universe $u = 1$.

Cell 21 is in universe $u = 5$ and is a 13×13 rectangular lattice (lat = 1). Each lattice element has an index, $[i,j,k]$, which corresponds to the x, y, z axes because each lattice box surface, 73, is an RPP (see surfaces, below) type surface. Whereas the lattice box surface 73 is centered exactly at the origin, the lattice element at the origin has i,j,k indices = $[0,0,0]$. The i and j indices go from -6 to $+6$, specifying 13 lattice elements in the x- and y-directions; the k indices are from 0 to 0, indicating only one lattice element in the z-direction. The lattice elements are filled with universes 5, 1, and 8. Note how $u = 5$ can be filled with itself—namely $u = 5$—which, in this case, is a lattice box of water. Note also how the order of the filling elements in the lattice lines of cell 21 looks like the upside-down version of the geometry illustrated in Fig. 3.1 and shown in detail in Fig. 3.2.

Cells 16 and 17 describe universe $u = 8$, which also fills the lattice of universe $u = 5$. Universe 8 is depleted uranium inside surface 76 and extends to infinity outside surface 76. Although universes 1 and 8 go out to infinity, they are made finite by being contained in the lattice boxes of universe 5.

Cell 31 is in universe $u = 0$, which is the default universe and is filled with the lattice of universe 5. Cell 31 is inside RPP surface 74, which is exactly on the outside boundary of lattice boxes of universe 5. This bounding RPP can be smaller than the lattice of its filling universe but not larger because then there would be undefined space in the geometry. Cell 32 is the $u = 0$ cylindrical water tank outside the fuel assembly box. Cell 99 is the world outside the water tank, which extends to infinity but has importance imp:n = 0, which causes any neutrons that enter cell 99 to be terminated and no longer tracked in the problem.

The surfaces are RPP and RCC right parallelepiped and right circular cylinders, respectively. The RPP entries are $-x$ $+x$ $-y$ $+y$ $-z$ $+z$. The RCC entries are the coordinates of the base center, x, y, z, followed by the coordinates of the height vector in the x, y, z directions, followed by the cylinder radius.

3.1.3   Other Data: Sources, Materials, Tallies, and More

After the third blank line follows the non-geometry information of the problem.

```
SDEF PAR=SF  pos=0 0 0  ext=d61  axs=0 0 1  rad=fcel=d50     cel=d63
si61 -150 150
sp61    0   1
si63 L    (11<21[-5:2  -6:-6 0:0]<31)
          (16<21[ 3:6  -6:-6 0:0]<31)
          (11<21[-6:6  -5:6  0:0]<31)
sp63      2.68 7R   5.80 3R   2.68 155R
DS50 S      51 7R      52 3R     51 155R
si51    0 .4840
sp51 -21 1
si52    0 .700
sp52 -21 1
```

SDEF is the source definition line that specifies the source as spontaneous fission, PAR = SF, in a cylinder with a base at the origin, POS = 0 0 0, and axis in the z-direction, AXS = 0 0 1. The height is sampled linearly according to distribution EXT = D61, which goes from −150 to +150 cm. All 168 active source cells are listed by i,j,k indices in the cell distribution (d63): cell 11 ($u = 1$) eight times for $-5 \le i \le 2, j = -6$; then cell 16 ($u = 8$) four times for $3 \le 6, j = -6$; and then cell 11 ($u = 1$) for 156 times for $-6 \le i \le 6$ and $-5 \le j \le 6$. The source strengths (SP63) are 2.68 fissions/s for cell 11 and 5.8 fissions/s for cell 16; these are calculated from the spontaneous fission yields times component nuclide mass fraction times cell mass (from Print Tables 38, 40, and 50 in the OUTP file). The radii are dependent on the cell (dependent distribution DS50) and sampled uniform in the area using distribution D51 for cell 11 (radius 0.4840) and distribution D52 for cell 16 (radius 0.7000).

The full print (print) is turned on. An MCTAL file is written for post-processing (PRDMP 2J 1). The source strength totaled over all fuel pins is 1 neutron (WGT not specified on SDEF) but it is sampled with 10,000 source histories (nps 1E4). The number of source histories sampled affects only statistical convergence and has no physical meaning.

Material 102 in cell 16 is depleted uranium oxide (m102) with uranium atom fractions 0.3 ^{235}U and 99.7 ^{238}U. Material 201 consists of five zirconium isotopes, ^{90}Zr, ^{91}Zr, ^{92}Zr, ^{94}Zr, and ^{96}Zr, with mass fractions 0.505239, 0.110180, 0.168413, 0.170672, and 0.027496. Material 313 is uranium dioxide, UO_2, 3.17 at% or 3.13 wt% enriched in ^{235}U. Material 101 is light water. MT101 specifies the light water molecular $S(\alpha,\beta)$ treatment, which should be provided when available for all light mass materials in neutron calculations. All cross-sections are specified as ".80c", which tells the code to use the ENDFVII revision 1 version of the data libraries, if available, or the ENDFVII.1.

```
FC4 Induced fission neutrons in enriched and depleted fuel pins
F4:n   (11<(21[-5:2  -6:-6 0:0])<31) (11<(21[-6:6  -5:6  0:0])<31)
       (16<(21[ 3:6  -6:-6 0:0])<31)
FQ4 F M
FM4  (-1 102 -6 -7) (-1 313 -6 -7)
sd4 1 2R
```

Tally 4 is described in the tally comment (FC4) and tallies the number of induced fissions in the 168 fuel pins (F4:n) printed with multiplier bins horizontal and fuel pin cells vertical (FQ4). The two multiplier bins (FM4) multiply the flux by $\nu\sigma_f$ to convert flux to fission neutrons/volume, respectively, for enriched fuel (material 102) and for depleted uranium (material 313). The -1 multiplies by atom density of the material; the reactions -6 and -7 are the microscopic fission cross-section, and the ν, fission multiplicity. SD4 overrides the volume division to provide final results in fission neutrons per source neutron, which is also equivalent to k_{eff} in near-critical calculations.

3.1.4   MCNP Output

Here, we comment on the output and the results obtained from running the MCNP code with the input file of Example 3.1. Some of the output is optional and requires the PRINT card to be in the input file.

Optional Print Table 38 reports the fission multiplicity data. Because there is no FMULT card in the input of Example 3.1, the default values are used. The default Gaussian widths are from Santi et al. [1]. Fission multiplicity is described quite well with a good example in the MCNP manual, Ref. [2] of Chap. 2, Section 3.3.3.8 FMULT. Print Table 38 is displayed later in Fig. 5.1.

Print Table 100 reports on the cross-sections used in the problem and should always be consulted to ensure that good data are being used.

Print table 110 provides information on the birth of the first 50 particles so that the user can verify the source is as desired.

The problem summary table provides the first results in terms of neutron creation and neutron loss in the problem shown in Fig. 3.3:

For the creation (left side), we can read that prompt fission neutron production is 4.1071E+00 and delayed neutron production is 2.9445E−02 per source neutron. (The weight represents the real number of neutrons divided into an arbitrary number of tracks depending on the setup of the problem.) For neutron loss (right side), the major loss is due to the neutron capture, 3.2721E+00, per source neutron.

The same table reports a key result of the problem—the net multiplication (3.5233 with relative uncertainty of 0.0133).

The results of the F4 tally, as fission per source neutron, for the three cells are reported in Fig. 3.4.

```
1problem summary                   source particle weight for summary table normalization =     19878.00

+    run terminated when     10000  particle histories were done.

====>   0.42 M histories/hr    (based on wall-clock time in mcrun)          04/02/19 14:07:42

   Fuel Assembly in water Tank      probid =  04/02/19 14:06:13

neutron creation    tracks      weight         energy          neutron loss         tracks      weight          energy
                             (per source particle)                                           (per source particle)

source              19878   1.0000E+00   1.6921E+00       escape                8639   1.7887E-01   1.9043E-01
nucl. interaction       0   0.           0.              energy cutoff            0   0.           0.
particle decay          0   0.           0.              time cutoff              0   0.           0.
weight window           0   0.           0.              weight window            0   0.           0.
cell importance         0   0.           0.              cell importance          0   0.           0.
weight cutoff           0   1.5497E+00   4.4990E-02       weight cutoff        132041   1.5528E+00   4.8313E-02
e or t importance       0   0.           0.              e or t importance        0   0.           0.
dxtran                  0   0.           0.              dxtran                   0   0.           0.
forced collisions       0   0.           0.              forced collisions        0   0.           0.
exp. transform          0   0.           0.              exp. transform           0   0.           0.
upscattering            0   0.           2.4948E-06       downscattering           0   3.2721E+00   9.4166E+00
photonuclear            0   0.           0.              capture                  0   3.4017E-03   1.1006E-01
(n,xn)                333   6.8336E-03   4.7447E-03       loss to (n,xn)         166   1.6859E+00   2.9063E-02
prompt fission     202269   4.1071E+00   8.3226E+00       loss to fission      83071   1.6859E+00   2.8453E-01
delayed fission      1437   2.9445E-02   1.4551E-02       nucl. interaction        0   0.           0.
prompt photofis         0   0.           0.              particle decay           0   0.           0.
tabular boundary        0   0.           0.              tabular boundary         0   0.           0.
tabular sampling        0   0.           0.              elastic scatter          0   0.           0.
   total            223917   6.6931E+00   1.0079E+01         total             223917   6.6931E+00   1.0079E+01

                                                         average time of (shakes)           cutoffs
number of neutrons banked              134582            escape        2.7375E+07            tco   1.0000E+33
neutron tracks per source particle  1.1265E+01           capture       3.3248E+07            eco   0.0000E+00
neutron collisions per source particle 1.2011E+03        capture or escape 3.2944E+07        wc1  -5.0000E-01
total neutron collisions            23376450             any termination 2.7946E+07          wc2  -2.5000E-01
net multiplication    3.5233E+00 0.0133

computer time so far in this run     1.47 minutes        maximum number ever in bank       48
computer time in mcrun               1.43 minutes        bank overflows to backup file      0
source particles per minute        6.9777E+03
random numbers generated            223658553            most random numbers used was      5187  in history

                                                                                     475862  in history
```

Fig. 3.3 The MCNP problem summary is provided in every MCNP output file regardless of PRINT card options. It is a balance of creation and loss. The tracks are MCNP histories and relate to statistical sampling with no physical meaning. The weight relates to physical particles per source particle. The energy is the average energy times the weight in MeV per source particle; the average energy of a particular event is then the energy divided by weight

```
1tally        4        nps =        10000
+                                    Induced fission neutrons in enriched and depleted fuel pins
           tally type 4     track length estimate of particle flux.
           particle(s): neutrons
           cell   a is (11<21[-5 -6 0] 21[-4 -6 0] 21[-3 -6 0] 21[-2 -6 0] 21[-1 -6 0] 21[0 -6 0] 21[1 -6 0] 21[2 -6 0]<31)
           cell   b is (11<21[-6 -5 0] 21[-5 -5 0] 21[-4 -5 0] 21[-3 -5 0] 21[-2 -5 0] 21[-1 -5 0] 21[0 -5 0] 21[1 -5 0] 21[2 -5 0] 21[
 3 -5 0] 21[4 -5 0] 21[5 -5 0] 21[6 -5 0] 21[-6 -4 0] 21[-5 -4 0] 21[-4 -4 0] 21[-3 -4 0] 21[-2 -4 0] 21[-1 -4 0] 21[0 -4 0] 21[1 -4 0]
 21[2 -4 0] 21[3 -4 0] 21[4 -4 0] 21[5 -4 0] 21[6 -4 0] 21[-6 -3 0] 21[-5 -3 0] 21[-4 -3 0] 21[-3 -3 0] 21[-2 -3 0] 21[-1 -3 0] 21[0
-3 0] 21[1 -3 0] 21[2 -3 0] 21[3 -3 0] 21[4 -3 0] 21[5 -3 0] 21[6 -3 0] 21[-6 -2 0] 21[-5 -2 0] 21[-4 -2 0] 21[-3 -2 0] 21[-2 -2 0]
21[-1 -2 0] 21[0 -2 0] 21[1 -2 0] 21[2 -2 0] 21[3 -2 0] 21[4 -2 0] 21[5 -2 0] 21[6 -2 0] 21[-6 -1 0] 21[-5 -1 0] 21[-4 -1 0] 21[-3 -1
0] 21[-2 -1 0] 21[-1 -1 0] 21[0 -1 0] 21[1 -1 0] 21[2 -1 0] 21[3 -1 0] 21[4 -1 0] 21[5 -1 0] 21[6 -1 0] 21[-6 0 0] 21[-5 0 0] 21[-4 0
0] 21[-3 0 0] 21[-2 0 0] 21[-1 0 0] 21[0 0 0] 21[1 0 0] 21[2 0 0] 21[3 0 0] 21[4 0 0] 21[5 0 0] 21[6 0 0] 21[-6 1 0] 21[-5 1 0] 21[-4
1 0] 21[-3 1 0] 21[-2 1 0] 21[-1 1 0] 21[0 1 0] 21[1 1 0] 21[2 1 0] 21[3 1 0] 21[4 1 0] 21[5 1 0] 21[6 1 0] 21[-6 2 0] 21[-5 2 0] 21[
-4 2 0] 21[-3 2 0] 21[-2 2 0] 21[-1 2 0] 21[0 2 0] 21[1 2 0] 21[2 2 0] 21[3 2 0] 21[4 2 0] 21[5 2 0] 21[6 2 0] 21[-6 3 0] 21[-5 3 0]
21[-4 3 0] 21[-3 3 0] 21[-2 3 0] 21[-1 3 0] 21[0 3 0] 21[1 3 0] 21[2 3 0] 21[3 3 0] 21[4 3 0] 21[5 3 0] 21[6 3 0] 21[-6 4 0] 21[-5 4
0] 21[-4 4 0] 21[-3 4 0] 21[-2 4 0] 21[-1 4 0] 21[0 4 0] 21[1 4 0] 21[2 4 0] 21[3 4 0] 21[4 4 0] 21[5 4 0] 21[6 4 0] 21[-6 5 0] 21[-5
5 0] 21[-4 5 0] 21[-3 5 0] 21[-2 5 0] 21[-1 5 0] 21[0 5 0] 21[1 5 0] 21[2 5 0] 21[3 5 0] 21[4 5 0] 21[5 5 0] 21[6 5 0] 21[-6 6 0] 21[
-5 6 0] 21[-4 6 0] 21[-3 6 0] 21[-2 6 0] 21[-1 6 0] 21[0 6 0] 21[1 6 0] 21[2 6 0] 21[3 6 0] 21[4 6 0] 21[5 6 0] 21[6 6 0]<31)
           cell   c is (16<21[3 -6 0] 21[4 -6 0] 21[5 -6 0] 21[6 -6 0]<31)
           multiplier bin  1:   -1.00000E+00          102           -6          -7
           multiplier bin  2:   -1.00000E+00          313           -6          -7

           volumes
                   cell:          a             b             c
                         1.00000E+00   1.00000E+00   1.00000E+00

      mult bin:               1                        2
         cell
           a           3.14368E-02 0.0217    2.37809E-01 0.0231
           b           5.70713E-01 0.0179    3.87062E+00 0.0182
           c           3.65714E-02 0.0246    3.19811E-01 0.0265
```

Fig. 3.4 Tally F4 results from Example 3.1. The tally is labeled with the description from the FC4 tally comment card. The lattice elements of the three cells, *a*, *b*, *c*, are specifically listed. Then the results for each multiplier bin are tabulated. The FQ4 card makes this tabulation more readable by listing the multiplier bins horizontally and the cells vertically

Table 3.1 Tally F4 tabulated results repeated in a table format

	Multiplier: (−1 102 −6 −7)		Multiplier: (−1 313 −6 −7)	
Cell	Value	Relative error	Value	Relative error
11 < (21[−5:2 −6:−6 0:0]) < 31	3.14E−02	0.0217	2.38E−01	0.0231
11 < (21[−6:6 −5:6 0:0]) < 31	5.71E−01	0.0179	3.87E+00	0.0182
16 < (21[3:6 −6:−6 0:0]) < 31	3.66E−02	0.0246	3.20E−01	0.0265

In Table 3.1, we have extracted the information to be more easily read.

The range of lattices for cells a, b, and c are listed in the first column. The multiplier bins are the number of induced fissions per gram in material 102 (columns 2 and 3) and material 313 (columns 4 and 5). The relative errors are the standard deviation divided by the tally value to give a fractional error between 0.0 and 1.0. Only the enriched uranium material 313 results are valid for the enriched uranium cell 11 and only the depleted uranium material 102 results are valid for the depleted uranium cell 16.

This table shows us that the fission-neutron production rate is about 10 times higher in the low-enriched uranium pins (cell 11) compared with the depleted uranium pins (cell 16), as expected, and that the statistical uncertainty on the results is between 1.8% and 2.7%. The 2.38E−01 induced fissions in the eight low-enriched pins on the edge of the assembly is less than the 3.87E+00 fissions averaged over the other 156 low-enriched pins and more than the 3.66E−02 fissions in the depleted uranium pins. The F4 track length tally of fission neutron production is thus 2.38E-01 + 3.87E+00 + 3.66E-02 = 4.474E+00. The net multiplication is the neutron production less the neutron source, or 3.474E+00. This track-length estimate agrees within statistics with the net multiplication of 3.5233 in Figure 3.3.

3.2   Coincidence Counter with F4 and F8 Tallies for Coincidence and Multiplicity Counting Rates

3.2.1   Description

Example 3.2 shows how to use the MCNP code to calculate Singles and Doubles (coincidence) counting rates in a coincidence counter. The theory and operation of coincidence counting are described in the PANDA manual [2] and the input file describes a crude model of an HLNC2 detector taken from page 499 of the same reference. This example illustrates how the Singles, Doubles, Triples, etc., rates can be calculated from the F4 tally using the point model or directly from the F8 tally without relying on the point model assumptions.

The different parts of the input file will be described and then the output file and tally results will be explained.

Example 3.2 Coincidence Counter for Multiplicity Counting

```
Crude model of HLNC2 based on p499 of PANDA Manual
c    Cells
c    sample volume
100 94 -2.5 -100 imp:n=1
c    Cavity
1    0 -1 100 imp:n=1
c    Bottom end plug poly
2    1 -0.96 -2 3  imp:n=1
c    Bottom end plug Al
3    13 -2.7 -3  imp:n=1
c    Bottom end plug Cd
4    48 -8.64 -4 2 imp:n=1
c    Top end plug poly
5    1 -0.96 -5 6  imp:n=1
c    Top end plug Al
6    13 -2.7 -6 imp:n=1
c    Top end plug Cd
7    48 -8.64 -7 5 6  imp:n=1
c    Cd liner
8    48 -8.64 -11 1 4 7 imp:n=1
c    Poly moderator with external cutouts and holes for #He tubes
9    1 -0.96 (-8  : -9  : -10 ) 11 12 #11 #12 #13 #14 #15 #16 #17 #18
                              #19 #20 #21 #21 #22 #23 #24 #25 #26
#27 imp:n=1
c    18 3He detectors 4 atm 3He only (no walls, air gap, top & bot-
tom end spaces)
10   3  -4.99e-4 -12  imp:n=1
11   Like 10 but TRCL=(-0.730      4.138   0 )
12   Like 10 but TRCL=(-2.831      7.778   0 )
13   Like 10 but TRCL=(-6.050     10.479   0 )
14   Like 10 but TRCL=(-9.999     11.916   0 )
15   Like 10 but TRCL=(-14.201    11.916   0 )
16   Like 10 but TRCL=(-18.150    10.479   0 )
17   Like 10 but TRCL=(-21.369     7.778   0 )
18   Like 10 but TRCL=(-23.470     4.138   0 )
19   Like 10 but TRCL=(-24.200     0.000   0 )
20   Like 10 but TRCL=(-23.470    -4.138   0 )
21   Like 10 but TRCL=(-21.369    -7.778   0 )
22   Like 10 but TRCL=(-18.150   -10.479   0 )
23   Like 10 but TRCL=(-14.201   -11.916   0 )
24   Like 10 but TRCL=(-9.999    -11.916   0 )
25   Like 10 but TRCL=(-6.050    -10.479   0 )
26   Like 10 but TRCL=(-2.831     -7.778   0 )
27   Like 10 but TRCL=(-0.730     -4.138   0 )
c    Outside vacuum
```

```
110   0   -999 8 9 10 imp:n=1
c    Exterior world
120   0 999 imp:n=0

c    surfaces
1    rcc 0     0 -20.3412 0 0 40.6824 8.75    $ cavity
2    rcc 0     0 -28.8     0 0 8.3      8.75    $ Bottom end plug poly
3    rcc .0     0 -26.7     0 0 6.2      6.75    $ Bottom end plug Al
4    rcc 0     0 -28.8     0 0 8.4588   8.75    $ Bottom end plug Cd
5    rcc 0     0  20.5     0 0 8.3      8.75    $ Top end plug poly
6    rcc 0     0  20.5     0 0 6.2      6.75    $ Top end plug Al
7    rcc 0     0  20.3412 0 0 8.4588   8.75    $ Top end plug Cd
8    rcc 0     0 -28.8     0 0 16.8     16.9    $ bottom poly moderator
9    rcc 0     0 -12.      0 0 24       15      $ middle poly moderator
10   rcc 0     0  12       0 0 16.8     16.9    $ top poly moderator
11   rcc 0     0 -28.8     0 0 57.6     8.9088 $ Cd liner
12   rcc 12.1 0 -25.4     0 0 50.8     1.27    $ 3He tube
100  rcc 0     0 -10       0 0 20       5       $ Sample volume
999  rcc 0     0 -30       0 0 60       20

c    Materials
c    poly density 0.96 g/cm3
m1   1001 2 6012 1
mt1  poly.10T  $ specification of thermal S(a,b) scattering
c    Al
m13 13027 1
c    Cd
m48 48000 1
c 3He
m3   2003. 1
c plutonium oxide
m94  94240 0.2 94239 0.8 8016 2.0
phys:n   25 25 0 J J J 0 30 J J J 0 0
fmult    94239 method=5 data=3 shift=1 $ method=LLNL data=Ens.
shift=MCNPX
fmult    94240 method=5 data=3 shift=1 $
c    spontaneous fission source in sample volume
c    Source nuclide is chosen from possible spontaneous fission
nuclides present
sdef par=sf  ext=d1 axs = 0 0 1 rad=d2
si1   -10 10
sp1    0   1
si2    0   5
sp2   -21  1
c
c    sdef par=n  ext=d1 axs = 0 0 1 rad=d2 erg=d3
```

```
c      si1    -10 10
c      sp1     0   1
c      si2     0   5
c      sp2    -21  1
c energy cards for (alpha,n) source on Oxygen
c      si3 h 0.00E+00 2.00E-01 4.00E-01 6.00E-01 8.00E-01
c             1.00E+00 1.20E+00 1.40E+00 1.60E+00 1.80E+00
c             2.00E+00 2.20E+00 2.40E+00 2.60E+00 2.80E+00
c             3.00E+00 3.20E+00 3.40E+00 3.60E+00 3.80E+00
c             4.00E+00 4.20E+00 4.40E+00 4.60E+00 4.80E+00
c             5.00E+00 5.20E+00 5.40E+00 5.60E+00 5.80E+00
c             6.00E+00 6.20E+00 6.40E+00 6.60E+00 6.80E+00
c             7.00E+00 7.20E+00 7.40E+00 7.60E+00 7.80E+00
c             8.00E+00 8.20E+00 8.40E+00 8.60E+00 8.80E+00
c             9.00E+00 9.20E+00 9.40E+00 9.60E+00 9.80E+00
c             1.00E+01 1.02E+01 1.04E+01 1.06E+01 1.08E+01
c             1.10E+01 1.12E+01
c      sc3   PuO2 source (a,n) only
c      sp3 d 0.00E+00 1.40E-02 2.02E-02 1.93E-02 1.67E-02
c             1.89E-02 2.52E-02 3.42E-02 4.54E-02 5.84E-02
c             7.62E-02 9.14E-02 1.05E-01 1.06E-01 9.80E-02
c             8.36E-02 6.64E-02 5.08E-02 3.39E-02 1.88E-02
c             9.46E-03 3.63E-03 1.39E-03 9.54E-04 7.04E-04
c             5.53E-04 3.74E-04 2.08E-04 3.57E-05 3.40E-06
c             3.57E-08 0.00E+00 0.00E+00 0.00E+00 0.00E+00
c             0.00E+00 0.00E+00 0.00E+00 0.00E+00 0.00E+00
c             0.00E+00 0.00E+00 0.00E+00 0.00E+00 0.00E+00
c             0.00E+00 0.00E+00 0.00E+00 0.00E+00 0.00E+00
c             0.00E+00 0.00E+00 0.00E+00 0.00E+00 0.00E+00
c             0.00E+00 0.00E+00
c tallies (sum of all 3He detectors)
c
c He-3 Interactions
fc4 He-3 interactions from track length tally
f4:n (10 11 12 13 14 15 16 17 18 19 20 21 22 23 24 25 26 27)
fm4  -1.0 3 103 $integral of flux*(n,p)reaction of material 3 * 1.0
*atom density
sd4   1 $ divide by 1 instead of cell volume to get total reac-
tion rate
t4    450 6850 20e8 T  $ time bins to give the coincidence gate
fraction
c
c Coincidence tallies based on capture in 3He
c
fc8 Ungated Coincidence tally
f8:n (10 11 12 13 14 15 16 17 18 19 20 21 22 23 24 25 26 27)
```

```
ft8   cap 2003
c
fc18 Gated Coincidence tally
f18:n (10 11 12 13 14 15 16 17 18 19 20 21 22 23 24 25 26 27)
ft18   cap 2003 gate 450 6400 $ 1us=10 shakes --> 4.5 us predelay
and 64 us gate
c
FC108 Full gate to compare moments with ungated F8
F108:n (10 16I 27)
FT108   CAP  2003 gate 0 1e20
c
FC116  Energy deposition tally to demonstrate FT118 CAP EDEP
F116:n (10 16I 27)
c
FC118  Energy deposition capture tally
F118:n  (10 16I 27)
FT118   CAP  EDEP 116 .5
print -30 -50 -70 -140 -162
nps 1e6
```

The geometry is a very simplified model of the HLNC2 detector, as shown in the PANDA manual p. 499 [2]. A vertical and horizontal cross-section of the geometry is shown in Fig. 3.5.

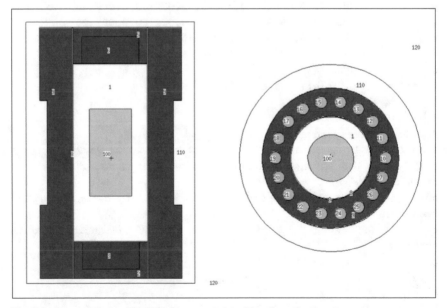

Fig. 3.5 Cross-sections of HLNC2 detector model with cell numbers

The item and the ^{3}He detectors are modeled with no structure (no detector walls, no deadspaces, no quench gas, etc.). No allowance is made for the presence of the electronics or any external structures, such as the floor. The cadmium liners of the cavity and endplugs are included. The ^{3}He gas density is based on pure ^{3}He at 4 atm pressure and 20 °C.

The geometry is built of simple macrobodies. The ^{3}He detectors are based on 1 macrobody that is copied by 17 TRCL translations, thus the density, diameter, and length of all detectors are controlled by single entries, ensuring consistency and making them easy to change. (The translations were calculated with Excel and cut and pasted into the file.) The item is of 3.9 kg of plutonium dioxide at a (powder) density of 2.5 g/cm^3.

3.2.2   Materials

The plutonium composition is 20% ^{240}Pu and 80% ^{239}Pu. All cross-sections were entered without any library identifiers, which causes the cross-sections to be chosen from the XSDIR file entries. Particular versions can be chosen if desired. The specification of carbon in polyethylene as ZAID = 6012 causes MCNP to use the 1979 ^{12}C evaluation unless ENDF/B-VIII is available. ENDF/B-VI and ENDF/B-VII data evaluations use ZAID = 6000. Note that natural isotopic composition is defined for cadmium. Elemental cadmium (material 48,000) is a 1980 evaluation. Modern evaluations require specifying the isotopic composition of cadmium. The remaining cross-sections used are .80c ENDF/B_VII.1 cross-sections. *One of the most common MCNP error is to not examine Print Table 101 in the output to ensure that the cross-sections MCNP is using are the ones intended.* Note nucleardata.lanl.gov is a useful resource for nuclear data libraries.

3.2.3   Source

This item produces both spontaneous fission and (α,n) neutrons. In the field of nondestructive measurements for safeguards, the ratio of (α,n) neutrons to spontaneous fission neutrons is called α. The counting rates from each source are completely separate with no interactions. *NOTE: This calculation—either with the point model or with the F8 tally—does not include the "accidental coincidence rate," which is, in any case, subtracted to produce the measured value. This is a significant advantage for the precision of the calculation compared with the precision of an actual measurement. (If it is desired to model the experimental case with accidentals, it is necessary to use the PTRAC capability (MCNP manual Section 3.3.7.2.4) to produce events and then combine them into a simulated pulse train.)* The calculation can be done either in one step, with a combined spontaneous fission/(α,n) source, or in two separate steps, in which the count rates from each source are calculated separately. The second method is chosen here because, although it requires two calculations rather than one, it does mean that, with these two runs, results can be produced for any value of α rather than requiring one run for every α value.

The source is chosen to be spontaneous fission (PAR = sf) or (α,n) over the height and radius of the item. (In the example file above (Example 3.2), the (α,n) source energy distribution is commented out.) The probability in the vertical direction is uniform, but the probability in the radial direction uses a built-in probability distribution in the source description (sp2 -21 a), which gives a power law $\alpha |r|^a$ for the probability so that the volume of the item is uniformly sampled. The power law power of $a = 1$ is used because the differential cylindrical volume is $dV = 2\pi h r^1 dr$, where the radius is sampled to the first power.

For spontaneous fission, the starting source event is chosen from all possible nuclides present in the cell that can undergo spontaneous fission, properly weighted with their abundance and specific spontaneous fission probability (predominantly ^{240}Pu in this case). When $PAR = sf$ is used, the tallies are normalized to the number of starting neutrons in the same way as $PAR = n$. If $PAR = -sf$ is used, the tallies are normalized to the number of starting fissions. The difference between the results for the two options is the average nu-bar of the source. The F8 tally results are always presented normalized in both ways. The nps value that controls the length of the run (in both options) will be in fissions so that the number of starting neutrons will be nps*nu-bar. In the (α,n) source case, single neutrons are started with the energy distribution given by the si3 and sp3 cards from the d3 distribution. In this case, the number of neutrons is simply nps. The energy spectrum of the (α,n) source used here is given as an example. It is not well documented and should be used with care.

3.2.4 Tallies

The F4 tally calculates the reaction rate in the sum of all detector regions by integrating the track length estimate of the flux as a function of energy together with the (n,p) reaction cross-section of ^{3}He (type 103; see Chapter 2 Table 2.1 or Table 3-104 of the MCNP6 manual) controlled by the entries on the FM card. The first entry on the FM card requests that the integral be multiplied by 1 and the negative sign requests that the integral be multiplied also by the atom density of material 3. The result would normally be divided by the volume of the (sum of the) cells to give the reaction rate per unit volume, but the SD4 card sets the cell volume to one so that we obtain the total reaction rate over all of the cells. There is a time bin specified on the t4 card that sets a time bin that starts at the predelay and ends at the predelay + gate. The times are in shakes. This result will be used to obtain the coincidence gate fraction.

The F8 tally gives the ungated coincidence tally for the detector for each single history (including multiplication). The sum of the detectors is specified as for the F4 tally. The FT8 card requests that a pulse is considered to be created when there is a reaction with ^{3}He (ZAID 2003) in the detector volume. The tally makes a histogram of the number of created pulses for each history. At the end of the run, the factorial moments of this histogram are calculated. The results of this tally are presented with the other tallies toward the end of the output in a much more readable form in Print Table 118, just after the information about the source multiplicity in Print Tables 117 and 115. In this example, the F8 tally is ungated; no coincidence gate is applied. This tally gives us the Singles rate and the *ungated* Doubles and Triples, etc.

The F18 tally uses a coincidence gate based on the parameters predelay and gate following the "gate" keyword, in shakes

```
ft18  cap [-21] [-12] 2003 gate 450 6400
```

which extends from the predelay to the predelay + gate. *IMPORTANT NOTE: This tally gives us the observed Doubles and Triples, etc., but the first moment is the Doubles—not the Singles—as in the case of the ungated tally. In this case, the default values of mc (maximum number of captures = 21) and m0 (maximum number of moments = 12) are used and can be omitted. For highly multiplying items, these can be increased.*

3.2.5   Warning Messages

Warning messages indicate that there may be a user error and all should be carefully examined.

```
Warning: physics models disabled:
warning.  Memory reduction option specified, physics models disabled.
```

From Print Table 101 in the output file, the cross-section data tables extend from 0 to between 20 and 25.5 MeV. The default for neutron-only problems is not to use physics models for neutrons above this energy range but instead to use data at the highest energy in the data table. All code memory requirements for models are eliminated. Because this is a fission energy range calculation, the warning is not a concern.

```
Warning: tallies beyond last bin
 warning.  last time bin of tally      4 is less than time cutoff.
```

The MCNP code warns when the input is processed that the last time bin of tally 4 is less than the problem time cutoff (infinity). Later, during the particle transport, the message

```
warning. tally not scored beyond last time bin.
nps =     4252  tal =  4     tme = 2.0597E+09
```

indicates that statistical history 4252 could not score because it was past the last time bin. At the end of the output file, all scores beyond the last bin are summarized.

```
1     some tally scores were not made for various reasons:
                        beyond last bin        not in
            tally       user  segment    mult   angle   energy    time
                4           0        0       0       0        0     253
Warning: unnormalized material fractions
 warning.   2 materials had unnormalized fractions. print table 40.
```

The above warning is issued when the material fractions in the input do not add to either one or to the density on the cell card. (The code does renormalize the material fraction to 1.0.) In this calculation, the following specification of polyethylene is deliberate but not normalized.

```
m1   1001 2 6012 1
```

```
Warning: analog capture turned on
 warning.   Forced neutron analog capture for FT CAP or CGM option.
 comment.   unresolved resonances with variance reduction with
 f8 tally.
 warning.   Pulse-height tally variance reduction forced off. F8 tal-
lies not reliable.
```

Most correlated physics options, such as the CGM physics treatment or FT8 CAP coincident capture tallies, require analog capture with no particle weight adjustments for variance reduction see section 5.1.6. Whereas the unresolved resonance range probability table treatment—which is important for fission-energy problems—is turned on by default, The MCNP code recognizes that there will be automatic variance reduction whenever the probability tables overlap an energy cutoff boundary. Such an overlap will not occur in this calculation because there is no lower energy cutoff. Also, pulse-height tallies without the FT8 options use variance reduction by default, but that is automatically turned off by the FT8 option. Nonetheless, MCNP recognizes that there may be default variance reduction and issues these warnings.

3.2.6 Results

3.2.6.1 From the Point Model

From the point model, we have the Singles rate,

$$S = mG\varepsilon M v_s.$$

And the Doubles rate,

$$D = mGF_d\varepsilon M^2 v(v-1)/2,$$

where

m is the $^{240}\text{Pu}_{\text{eff}}$ mass
G is the number of fissions/s/g ^{240}Pu
F_d is the Doubles gate fraction
ε is the detection efficiency
M is the leakage multiplication
v_s is the mean number of neutrons per spontaneous fission (v_I for induced fission)
$v(v - 1)$ is the mean number of pairs per spontaneous fission including multiplication

Using Boehnel's approach [3], we usually write for the Doubles,

$$D = mG\varepsilon^2 F_d M^2 \frac{\overline{v_s(v_s-1)}}{2}\left[1+(M-1)(1+\alpha)K\right],$$

where

$$K = \frac{\overline{v_s v_i}\,\overline{(v_i-1)}}{\overline{(v_i-1)}\,\overline{v_s(v_s-1)}}.$$

K is taken as 2.166 (INCC [4]), and α is the ratio of (α,n) neutrons to spontaneous fission neutrons from the sample; however, it is more convenient to separate the spontaneous fission and (α,n) contributions as

$$D = mG\varepsilon^2 F_d M^2 \left[\frac{\overline{v_s(v_s-1)}}{2} + (M-1)\frac{\overline{v_s v_i}\,\overline{(v_i-1)}}{2\overline{(v_i-1)}}\right]$$
$$+ N_\alpha \varepsilon^2 F_d M^2 \left[(M-1)\frac{\overline{v_i(v_i-1)}}{2\overline{(v_i-1)}}\right],$$

where N_α is the (α,n) emission rate. This shows the similarity of the form between the coincidence rates caused by multiplication from spontaneous fission and (α,n) events. With the different contributions separated, we can now use the respective values for efficiency and gate fraction for the spontaneous fission case and the (α,n) case, although the values are usually very close.

α ($= N_\alpha/mGv_s$) can be calculated for materials with known stoichiometry and plutonium isotopic composition (p. 484 of Ref. [2]). For the plutonium oxide in the example, we have $\alpha = 0.2876$ for this rather idealized isotopic composition.

The quantities we need for the calculation (and where they come from) are shown in Table 3.2.

Extracts from the output file for the spontaneous fission cases are shown in Figs. 3.6, 3.7, 3.8, 3.9, and 3.10.

Table 3.2 Quantities needed to calculate coincidence and multiplicity counting rates and where to find them in the MCNP output file

Quantity	Location in output file	Value from spontaneous fission run	Value from (α,n) run
Mass of $^{240}Pu_{eff}$	Print Table 60 gives the PuO_2 mass for cell 100; Print Table 40 gives the Pu^{240} mass fraction for material 94	$3926.99 \times 0.177003 =$ 695.09 g (N.B. in general ^{238}Pu and ^{242}Pu need to be included to give $^{240}Pu_{eff}$)	Same
Multiplication	Problem Summary, five lines under the table "net multiplication" (equivalent to the "leakage multiplication" in [2])	1.1566	1.1624
v_s	Print Table 117	2.15454	n/a
	"spontaneous fission source multiplicity and moments"		
	"nu"		
	By number or by weight		
$v_s(v_s - 1)/2$	Print Table 117	1.89621	n/a
	"spontaneous fission source multiplicity and moments"		
	"nu(nu-1)/2!"		
	By number or by weight		
v_i	Print Table 117	3.10743	3.14732
	"induced fission source multiplicity and moments"		
	"nu"		
	By number or by weight		
$v_i(v_i - 1)/2$	Print Table 117	3.98444	4.0842
	"induced fission source multiplicity and moments"		
	"nu(nu-1)/2!"		
	By number or by weight		
Efficiency ε	Tally 4 total time bin, divided my multiplication	$0.202397/1.1566 = 0.1750$	$0.181203/1.1624 = 0.1559$
Gate fraction F_d	Tally 4 second time bin divided by total	$0.143773/0.202397 = 0.7104$	$0.128314/0.181203 = 0.7081$
N_α	$\alpha \times m \times G \times v_s$	n/a	$0.2876 \times 695.09 \times 473 \times 2.15454 = 203,726$

```
lcell volumes and masses                                                              print table 50

            cell     atom          gram          input         calculated                        reason volume
                     density       density       volume        volume           mass     pieces   not calculated

      1      100  1.66513E-02   2.50000E+00   0.00000E+00   1.57080E+03   3.92699E+03    1
      2        1  0.00000E+00   0.00000E+00   0.00000E+00   8.21447E+03   0.00000E+00    1
```

Fig. 3.6 Print Table 50

```
material
  number         component nuclide, mass fraction

      1              1001, 1.43814E-01   6012, 8.56186E-01
     13             13027, 1.00000E+00
     48             48000, 1.00000E+00
      3              2003, 1.00000E+00
     94             94240, 1.77003E-01  94239, 7.05059E-01 8016, 1.17938E-01
```

Fig. 3.7 Print Table 40

```
      Crude model of HLNC2 based on p499 of PANDA Manual     probid =  04/30/19 12:11:18

neutron creation    tracks      weight        energy        neutron loss      tracks      weight        energy
                                (per source particle)                                      (per source particle)

source             2154543   1.0000E+00   1.9336E+00        escape           1470548   6.8253E-01   7.4229E-01
nucl. interaction        0   0.           0.                energy cutoff          0   0.           0.
particle decay           0   0.           0.                time cutoff            0   0.           0.
weight window            0   0.           0.                weight window          0   0.           0.
cell importance          0   0.           0.                cell importance        0   0.           0.
weight cutoff            0   0.           0.                weight cutoff          0   0.           0.
e or t importance        0   0.           0.                e or t importance      0   0.           0.
dxtran                   0   0.           0.                dxtran                 0   0.           0.
forced collisions        0   0.           0.                forced collisions      0   0.           0.
exp. transform           0   0.           0.                exp. transform         0   0.           0.
upscattering             0   0.           1.0786E-07        downscattering         0   0.           1.6268E+00
photonuclear             0   0.           0.                capture          1020114   4.7347E-01   6.5515E-03
(n,xn)                 378   1.7544E-04   1.1810E-04        loss to (n,xn)       189   8.7722E-05   7.3068E-04
prompt fission      494291   2.2942E-01   4.7187E-01        loss to fission   159403   7.3985E-02   1.1928E-01
delayed fission       1042   4.8363E-04   2.3487E-04        nucl. interaction      0   0.           0.
prompt photofis          0   0.           0.                particle decay         0   0.           0.
tabular boundary         0   0.           0.                tabular boundary       0   0.           0.
tabular sampling         0   0.           0.                elastic scatter        0   0.           0.
    total          2650254   1.2301E+00   2.4058E+00            total        2650254   1.2301E+00   2.4058E+00

    number of neutrons banked                    1555188    average time of (shakes)              cutoffs
    neutron tracks per source particle      1.2301E+00      escape            4.0990E+05          tco   1.0000E+33
    neutron collisions per source particle  3.2031E+01      capture           7.9892E+05          eco   0.0000E+00
    total neutron collisions                  69011165      capture or escape 5.6923E+05          wc1   0.0000E+00
    net multiplication             1.1566E+00 0.0005        any termination   5.3496E+05          wc2   0.0000E+00
```

Fig. 3.8 Summary table showing multiplication for spontaneous fission case

For the singles, we have from spontaneous fission,
$S_{sf} = 695.09 \times 473 \times 0.175 \times 1.1566 \times 2.15454 = 143{,}370$ counts per second (cps).
And from (α,n),
$S_{\alpha} = 203{,}726 \times 0.1559 \times 1.1624 = 36{,}906$ cps.
Thus, the overall Singles rate is 180,276 cps.
For the Doubles, we have,
$D_{sf} = 695.09 \times 473 \times 0.175^2 \times 0.7104 \times 1.1566^2 \times [1.89621 + (1.1566 - 1) \times 2.15454 \times 3.98444/(3.10743 - 1)] = 24{,}246$ cps.
And from (α,n),
$D_{\alpha} = 203{,}726 \times 0.1559^2 \times 1.1624^2 \times 0.7081 \times (1.1624 - 1) \times 4.0842/(3.14732 - 1) = 1448$ cps.

```
1spontaneous fission source multiplicity and moments.                                    print table
117


          --------- by number ----------------          -------------- by weight --------------------------
                      fission    multiplicity                         fission    multiplicity
           fissions  neutrons      fraction        fissions          neutrons      fraction        error

  nu =  0     63324         0     6.33240E-02     2.93909E-02     0.00000E+00     6.33240E-02     0.0039
  nu =  1    231544    231544     2.31544E-01     1.07468E-01     1.07468E-01     2.31544E-01     0.0020
  nu =  2    333750    667500     3.33750E-01     1.54905E-01     3.09810E-01     3.33750E-01     0.0016
  nu =  3    252249    756747     2.52249E-01     1.17078E-01     3.51233E-01     2.52249E-01     0.0019
  nu =  4     98953    395812     9.89530E-02     4.59276E-02     1.83710E-01     9.89530E-02     0.0031
  nu =  5     18140     90700     1.81400E-02     8.41942E-03     4.20971E-02     1.81400E-02     0.0074
  nu =  6      2040     12240     2.04000E-03     9.46837E-04     5.68102E-03     2.04000E-03     0.0221

  total    1000000   2154543     1.00000E+00     4.64136E-01     1.00000E+00     1.00000E+00     0.0007


     factorial moments                by number                  by weight

                 nu        2.15454E+00 0.0005        2.15454E+00 0.0005
          nu(nu-1)/2!      1.89621E+00 0.0012        1.89621E+00 0.0012
       nu(nu-1)(nu-2)/3!   8.70261E-01 0.0022        8.70261E-01 0.0022
    nu(nu-1) .... (nu-3)/4!  2.20253E-01 0.0045      2.20253E-01 0.0045
    nu(nu-1) .... (nu-4)/5!  3.03800E-02 0.0099      3.03800E-02 0.0099
    nu(nu-1) .... (nu-5)/6!  2.04000E-03 0.0221      2.04000E-03 0.0221
```

Fig. 3.9   Print Table 117 for spontaneous fission case

```
1tally       4        nps =      1000000
+                                         He-3 interactions from track length tally
          tally type 4       track length estimate of particle flux.
          particle(s): neutrons
          cell   a is (10 11 12 13 14 15 16 17 18 19 20 21 22 23 24 25 26 27)

          volumes
                   cell:        a
                            1.00000E+00

  cell (10 11 12 13 14 15 16 17 18 19 20 21 22 23 24 25 26 27)
  multiplier bin:  -1.00000E+00           3            103
            time:       4.5000E+02              6.8500E+03           2.0000E+09              total
                        2.5236E-02 0.0021   1.43773E-01 0.0022   3.33877E-02 0.0048   2.02397E-01 0.0020
```

Fig. 3.10   Tally 4 for spontaneous fission case

Thus, the overall Doubles rate is 25,694 cps.

3.2.6.2   Rates Calculated without Point Model Assumptions

A better way to calculate the Doubles (coincidence rate) and the Triples, etc., rates is shown by using the F8 capture tally, which is not limited by the point model assumptions.

The ungated F8 results for spontaneous fission are shown in Fig. 3.11.

The factorial moments represent the (*ungated*) Singles, Doubles, Triples, Quads, Quints, etc., up to ninth order in this example. The values are given with two nor-malizations. In the factorial moments section, the "by number" column gives the tally per source event (spontaneous fission in this case) and the "by weight" column gives the tally per starting neutron. If we want to calculate the Singles rate, we can take the "3he" entry "by number" and multiply by the spontaneous fission rate of the item, or we can take the "by weight" entry and multiply by the starting neutron emission rate of the item. In other words, the ratio of the "by number" column to the "by weight" column is the starting multiplicity (2.1545 from Print Table 117 for the spontaneous fission run and 1.00 for the (α,n) run). (The 3he nomenclature reminds

```
1 neutron captures, moments and multiplicity distributions.  tally      8                    print table 118

weight normalization by source fission neutrons =        2154543
        cell   a is (10 11 12 13 14 15 16 17 18 19 20 21 22 23 24 25 26 27)

cell:      a

neutron captures on 3he
```

	histories	captures by number	captures by weight	multiplicity fractions by number	by weight	error
captures = 0	648298	0	0.00000E+00	6.48298E-01	3.00898E-01	0.0007
captures = 1	280871	280871	1.30362E-01	2.80871E-01	1.30362E-01	0.0016
captures = 2	58894	117788	5.46696E-02	5.88940E-02	2.73348E-02	0.0040
captures = 3	9796	29028	1.34729E-02	9.67600E-03	4.49098E-03	0.0101
captures = 4	1751	7004	3.25081E-03	1.75100E-03	8.12701E-04	0.0239
captures = 5	394	1970	9.14347E-04	3.94000E-04	1.82869E-04	0.0504
captures = 6	81	486	2.25570E-04	8.10000E-05	3.75950E-05	0.1111
captures = 7	23	161	7.47258E-05	2.30000E-05	1.06751E-05	0.2085
captures = 8	7	56	2.59916E-05	7.00000E-06	3.24895E-06	0.3780
captures = 9	5	45	2.08861E-05	5.00000E-06	2.32068E-06	0.4472
total	1000000	436342	2.01949E-01	1.00000E+00	4.64136E-01	0.0015

factorial moments	by number		by weight	
3he	4.37409E-01	0.0015	2.03017E-01	0.0015
3he(3he-1)/2!	1.04442E-01	0.0050	4.84752E-02	0.0050
3he(3he-1)(3he-2)/3!	2.38570E-02	0.0185	1.10729E-02	0.0185
3he(3he-1) .... (3he-3)/4!	6.86100E-03	0.0604	3.18443E-03	0.0604
3he(3he-1) .... (3he-4)/5!	2.38570E-03	0.1421	1.10696E-03	0.1421
3he(3he-1) .... (3he-5)/6!	8.58000E-04	0.2388	3.98228E-04	0.2388
3he(3he-1) .... (3he-6)/7!	2.59000E-04	0.3219	1.20211E-04	0.3219
3he(3he-1) .... (3he-7)/8!	5.20000E-05	0.3903	2.41350E-05	0.3903
3he(3he-1) .... (3he-8)/9!	5.00000E-06	0.4472	2.32068E-06	0.4472

Fig. 3.11 Ungated results for spontaneous fission case

```
1 neutron captures, moments and multiplicity distributions.  tally     18                    print table 118

weight normalization by source fission neutrons =        2154543
        cell   a is (10 11 12 13 14 15 16 17 18 19 20 21 22 23 24 25 26 27)

cell:      a

neutron captures on 3he

time gate: predelay =  4.5000E+02      gate width =  6.4000E+03
```

pulses in gate	histogram	occurrences by number	occurrences by weight	pulse fraction by number	by weight	error
captures = 0	373416	0	0.00000E+00	3.73416E-01	1.73316E-01	0.0014
captures = 1	55495	55495	2.57572E-02	5.54950E-02	2.57572E-02	0.0044
captures = 2	7186	14372	6.67056E-03	7.18600E-03	3.33528E-03	0.0128
captures = 3	1084	3252	1.50937E-03	1.08400E-03	5.03123E-04	0.0340
captures = 4	182	728	3.37891E-04	1.82000E-04	8.44727E-05	0.0840
captures = 5	36	180	8.35444E-05	3.60000E-05	1.67089E-05	0.1924
captures = 6	9	54	2.50633E-05	9.00000E-06	4.17722E-06	0.3685
captures = 7	1	7	3.24895E-06	1.00000E-06	4.64136E-07	1.0000
total	437409	74088	3.43869E-02	4.36342E-01	2.03017E-01	0.0045

factorial moments	by number		by weight	
n	7.40880E-02	0.0055	3.43869E-02	0.0054
n(n-1)/2!	1.20460E-02	0.0205	5.59098E-03	0.0205
n(n-1)(n-2)/3!	2.38700E-03	0.0654	1.10789E-03	0.0654
n(n-1)(n-2) ... (n-3)/4!	5.32000E-04	0.1556	2.46920E-04	0.1556
n(n-1)(n-2) ... (n-4)/5!	1.11000E-04	0.2856	5.15190E-05	0.2856
n(n-1)(n-2) ... (n-5)/6!	1.60000E-05	0.4841	7.42617E-06	0.4841
n(n-1)(n-2) ... (n-6)/7!	1.00000E-06	1.0000	4.64136E-07	1.0000

Fig. 3.12 Gated results for spontaneous fission case

us that we are looking at ^{3}He reactions independently of whether they occur in the electronic gate.) The fractional standard deviations of each column are the same. The statistical uncertainty on the Singles is 0.15%.

The gated (F18) results for spontaneous fission are shown in Fig. 3.12.

The results from the F18 tally represent the measured Doubles, Triples, Quads, Quints, etc., up to eighth order in this example. (The n, $n(n-1)/2!$. … nomenclature is used to distinguish these events in the gate from the actual ungated ^{3}He reactions.) We can use this tally to calculate the expected measured rates by multiplying the "by number" values by the spontaneous fission rate or by multiplying the "by weight" values by the spontaneous fission neutron emission rate.

We have also the ungated F8 for the (α,n) case shown in Fig. 3.13.

Here, the "by number" and "by weight" values are identical because we have exactly one neutron emitted per source event. The factorial moments table gives Singles, Doubles, Triples, etc., as before.

The gated tally for the (α,n) case is shown in Fig. 3.14.

These gated values correspond to the measured Doubles, Triples, Quads, etc. Regarding the ungated tally, the "by number" and "by weight" values are identical.

The results for both cases are combined into counting rates in Table 3.3. The "by number" values are used. The gate fractions (Doubles, Triples, etc.) can be calculated from the ratio of the gated tally to the ungated tally (making sure to use corresponding entries rather than the same row numbers).

Table 3.4 gives the sum of the count rates from spontaneous fission and (α,n) events.

We can compare the results calculated from the F4 tally with those from the gated F8 (F18 in this case). The difference between the results for the F18 and F4 methods is 0.3% for Singles and 1.6% for Doubles. One possible source of this difference is that the F4 tally is a track length estimate multiplied by the (n,p) cross section and the F8 tally is an independent collision estimate—two different statistical estimators. The difference between these two should become smaller as the

```
1 neutron captures, moments and multiplicity distributions.  tally        8                              print
table 118

weight normalization by source histories =      1000000
           cell   a is  (10 11 12 13 14 15 16 17 18 19 20 21 22 23 24 25 26 27)

cell:      a

neutron captures on 3he

                              captures          captures        multiplicity fractions
                  histories by number          by weight        by number        by weight        error

captures =  0       828469            0      0.00000E+00      8.28469E-01      8.28469E-01      0.0000
captures =  1       163620       163620      1.63620E-01      1.63620E-01      1.63620E-01      0.0023
captures =  2         6310        12620      1.26200E-02      6.31000E-03      6.31000E-03      0.0125
captures =  3         1288         3864      3.86400E-03      1.28800E-03      1.28800E-03      0.0278
captures =  4          248          992      9.92000E-04      2.48000E-04      2.48000E-04      0.0635
captures =  5           56          280      2.80000E-04      5.60000E-05      5.60000E-05      0.1336
captures =  6            5           30      3.00000E-05      5.00000E-06      5.00000E-06      0.4472
captures =  7            3           21      2.10000E-05      3.00000E-06      3.00000E-06      0.5773
captures =  9            1            9      9.00000E-06      1.00000E-06      1.00000E-06      1.0000

total             1000000     38029495      1.81436E-01      1.00000E+00      1.00000E+00      0.0023

   factorial moments              by number              by weight

               3he         1.81436E-01 0.0023      1.81436E-01 0.0023
          3he(3he-1)/2!    1.23960E-02 0.0153      1.23960E-02 0.0153
      3he(3he-1)(3he-2)/3! 3.12900E-03 0.0491      3.12900E-03 0.0491
   3he(3he-1) .... (3he-3)/4! 8.34000E-04 0.1791   8.34000E-04 0.1791
   3he(3he-1) .... (3he-4)/5! 2.75000E-04 0.4802   2.75000E-04 0.4802
   3he(3he-1) .... (3he-5)/6! 1.10000E-04 0.7718   1.10000E-04 0.7718
   3he(3he-1) .... (3he-6)/7! 3.90000E-05 0.9241   3.90000E-05 0.9241
   3he(3he-1) .... (3he-7)/8! 9.00000E-06 1.0000   9.00000E-06 1.0000
   3he(3he-1) .... (3he-8)/9! 1.00000E-06 1.0000   1.00000E-06 1.0000
```

Fig. 3.13 Ungated results for (α,n) case

statistical precision of the calculation improves. The remaining differences can be attributed to the limitations of the point model, primarily in this case, the use of the square of the average efficiency, ε^2, rather than the average of the efficiency squared $\overline{\varepsilon^2}$ to calculate the Doubles.

```
1 neutron captures, moments and multiplicity distributions.  tally      18
printtable 118

weight normalization by source fission neutrons =        1000000
            cell   a is (10 11 12 13 14 15 16 17 18 19 20 21 22 23 24 25 26 27)

cell:      a

neutron captures on 3he

time gate:  predelay =   4.5000E+02      gate width =  6.4000E+03

          pulses              occurrences       occurrences               pulse fraction
          in gate  histogram  by number         by weight         by number        by weight       error

captures =  0      174054         0            0.00000E+00      1.74054E-01      1.74054E-01       0.0022
captures =  1        6258       6258           6.25800E-03      6.25800E-03      6.25800E-03       0.0136
captures =  2         958       1916           1.91600E-03      9.58000E-04      9.58000E-04       0.0354
captures =  3         138        414           4.14000E-04      1.38000E-04      1.38000E-04       0.0928
captures =  4          23         92            9.20000E-05      2.30000E-05      2.30000E-05       0.2174
captures =  5           2         10            1.00000E-05      2.00000E-06      2.00000E-06       1.0000
captures =  6           2         12            1.20000E-05      2.00000E-06      2.00000E-06       1.0000
captures =  8           1          8            8.00000E-06      1.00000E-06      1.00000E-06       1.0000

   total           181436       8710           8.71000E-03      1.81436E-01      1.81436E-01       0.0136

      factorial moments              by number              by weight

              n                  8.71000E-03  0.0168     8.71000E-03  0.0168
          n(n-1)/2!              1.58800E-03  0.0636     1.58800E-03  0.0636
       n(n-1)(n-2)/3!            3.46000E-04  0.3035     3.46000E-04  0.3035
    n(n-1)(n-2) ... (n-3)/4!     1.33000E-04  0.7573     1.33000E-04  0.7573
    n(n-1)(n-2) ... (n-4)/5!     7.00000E-05  0.9718     7.00000E-05  0.9718
    n(n-1)(n-2) ... (n-5)/6!     3.00000E-05  1.0000     3.00000E-05  1.0000
    n(n-1)(n-2) ... (n-6)/7!     8.00000E-06  1.0000     8.00000E-06  1.0000
    n(n-1)(n-2) ... (n-7)/8!     1.00000E-06  1.0000     1.00000E-06  1.0000
```

Fig. 3.14 Gated results for (α,n) case

Table 3.3 F18 calculation values

Quantity	Spontaneous fission	(α,n) source
Source rate	695.09 × 473 × 2.15454 neutrons/s	203,759 neutrons/s
Singles tally (by weight)	2.03017E−01	1.81436E−01
Ungated Doubles tally	4.84752E−02	1.23960E−02
Ungated Triples tally	1.10729E−02	3.12900E−03
Ungated Quads tally	3.18443E−03	8.34000E−04
Doubles tally	3.43869E−02	8.71000E−03
Triples tally	5.59098E−03	1.58800E−03
Quads tally	1.10789E−03	3.46000E−04
Doubles gate fraction	0.7094	0.7026
Triples gate fraction	0.5049	0.5075
Quads gate fraction	0.3479	0.4149
Singles	1.438E+05	3.697E+04
Doubles	2.436E+04	1.775E+03
Triples	3.960E+03	3.236E+02
Quads	7.848E+02	7.050E+01

Table 3.4 Counting rates comparison from both methods

Quantity	F18 (sf + (α,n))	Point Model (sf + (α,n))	Difference %
Singles	1.808E+05	1.803E+05	−0.29
Doubles	2.613E+04	2.569E+04	−1.70
Triples	4.284E+03	n/a	n/a
Quads	8.55E+02	n/a	n/a

In addition, we can readily calculate the high-order coincidence rates from the F18 tally when we cannot from the F4 tally.

3.3   Gamma Pulse Height

3.3.1   Description and Input File

The NaI scintillation detector is typically used in laboratory and in-field to measure gamma ray spectra from nuclear materials. In Example 3.3, a 1.5 in. diameter × 2.0 in. height NaI detector is modeled in the MCNP code. The gamma source is a pellet of uranium oxide.

Example 3.3 Gamma Pulse Height Example with an NaI Scintillator

```
MCNP Simulation of NaI scintillation detector
c      diameter 1.5 " = 3.81 cm
c      length      2" = 5.08 cm
c
c    Cells
c
c     cel    mat      rho          geom
      10       1     -3.667        -1              $ NaI crystal
      20       2     -2.7          -2 1            $ Al around detector
      30       2     -2.7          -10             $ Al for PMT
      70      20     -10.34        -7              $ UO2 pellet
      40      10 -1.205e-3        -100
                        (2.1:10.2:2.3) (7.1:7.2:7.3)  $ Air
     100       0 100

c    Surfaces
c
1    RCC   0 0 0.05     0 0  5.13  1.905  $ NaI crystal
2    RCC   0 0 0        0 0  5.18  1.955  $ Al housing
7    RCC  -.5 0 -1.45   1 0  0     0.45   $ UO2 pellet
```

```
10    RCC   0 0 5.18   0 0  0.10   1.955  $ PMT behind the crystal
c
100     SO 100

c    Methods
c
nps 1e8
mode p e
imp:p 1 1 1 1 1 0  $
imp:e 1 1 1 1 1 0  $
c
c      Materials
c
M1      11000   1          $ Na    NaI, rho = -3.667 g/cm3
        53000   1          $ I
c
c
M2      13000   1          $ Al    Aluminum, rho = -2.70 g/cm3
c
M10      6000 -0.000124    $ C  Dry Air, rho = -1.205E-3 g/cm3
         7000 -0.755268    $ N
         8000 -0.231781    $ O
        18000 -0.012827    $ Ar
c
c
m20      92000 1           $ for gamma simulation, all U
                           $ rho=-10.34 g/cm3
         8000 2            $ O
c
c
c    Tally Section
c
FC8   NaI   Pulse Height Tally
F8:p 10                            $ Pulse Height Tally
E8 0 1.e-5  1e-3 1023I 0.512       $ energy bin, 0.5 keV bins
FT8 GEB 0.0 0.031196 4.6141        $ FWHM resolution
c
c
c
sdef PAR=p POS= 0 0 -1.45 RAD=D1 EXT=D2 AXS = 1 0 0  ERG=D3
c
SI1 0  0.45 $ radius, r
SP1 -21 1    $ sampling prop to r
c
```

```
SI2 -0.5 0.5 $ length
SP2 -21  0    $ default radius constant with x
C
C the # below indicates vertical formatting (MCNP6.2 manual 2.8.2)
#    SI3       SP3                     $ U-235
     L          D
   0.1438     7800                     $
   0.1634     3700                     $
   0.1857    43000
   0.2021      800
   0.2053     4000
```

3.3.2 Geometry

The photomultiplier behind the detector crystal was a model 1 mm thick aluminum disk. The pellet has a 0.45-cm radius and is 1 cm in length. The dimensions of the NaI detector are 3.81 cm in diameter and 5.08 cm in length.

3.3.3 Materials

For the photon transport problem of Example 3.3, the elemental ZAID is used for the material because the photons interact with the atomic electrons rather than the nucleus. For example, the uranium is specified as 92000, identifying the element independent of its isotopes. The specificity of that material is actually a ^{235}U isotope, which came from the definition of the gamma source (see *sdef* section of the example). Different ^{235}U enrichment would also affect the density of the pellet that was assumed as 10.34 g/cm³. In addition, the cross sections were left without any library identification. This choice causes the MCNP code to take the cross-section libraries as pointed in the XSDIR in use. The following file shows that, in the computer where the example ran, the plib = 84p and elib = 03e were used for photon and electron data, respectively.

3.3.4 Methods

Analog transport of both photons and electrons is used. In the MCNP code, the user has three options with the electrons: no transport of electrons, detailed transport of the electrons, and the TTB approximation (Sect. 2.2.5). If there is an "e" on the MODE card, as in the example, the full transport of the electrons is turned on. In this problem, there was no PHYS:P to control the transport of photons; the default parameter values are used.

The F8 tally is used for the pulse-height spectrum (see Sect. 2.4.5 of this guide). In the tally, the energy resolution of the detector is taken into account by means of the MCNP GEB special treatment that enables Gaussian energy broadening. The GEB function, as implemented in the MCNP code, is modeled by the following equation: $\text{FWHM}(E) = a + b\sqrt{E + cE^2}$, where E is the energy of the gamma ray in MeV (as we discussed in Sect. 2.4.4.2). It is worthwhile reminding the reader the structure of the energy bins for a pulse height that has a zero energy bin (see Sect. 2.4.5).

To determine the coefficients a, b, c, for this example, a set of measurements was performed in a laboratory using an actual NaI detector with the same crystal size as the one modeled in this example. The gamma energy lines 22, 59.5, 122, 356, and 662 keV from a set of calibration sources were used to extract the energy resolution of the detector, and the following coefficients were obtained: $a = 0$, $b = 0.31196$, and $c = 4.6141$.

The definition of the source is based on Table 7-1 of the PANDA manual [2]. For each gamma line used in the simulations the SP is the actual gamma/s/g of the ^{235}U. The MCNP code normalizes the SP probability for the source to 1.

3.3.5 Results

In Fig. 3.15, we present the pulse-height distribution as calculated. It is easy to identify the 186 keV—the most intense gamma ray contributing to the ^{235}U spectrum. Gamma-based methods to measure uranium enrichment use this gamma line. The intense peak at the energy around 100 keV can be observed here. This energy was not listed in the *sdef*. The peak we observe in the spectrum is actually a composite peak of the X-ray characteristic lines of uranium [2]. The low resolution of the NaI detector, modeled here with the GEB card, prevents us from distinguishing each characteristic X-ray line. A way to see the described effect is to run the input file with `FT8 GEB 0.0 0.031196 4.6141` commented out. Fig. 3.16 shows the spectrum obtained without GEB.

3.4 Active Neutron Example: Californium (Cf) Shuffler

3.4.1 Description and Input File

The Cf shuffler technique is well known for the measurement of nuclear material, especially uranium in waste [5]. A moveable Cf source is brought close to an item of interest in a detection chamber for a period of time and then withdrawn into a shielded location. The delayed neutrons from the induced fission are counted while the source is removed and used to quantify the mass of fissionable material in the item. The Cf shuffler assay detector is modeled in Example 3.4.

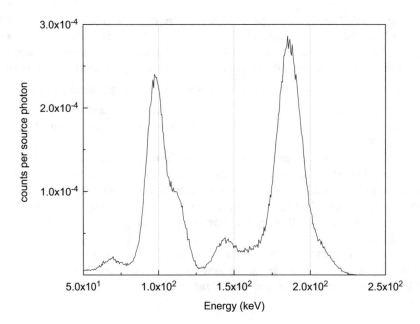

Fig. 3.15 MCNP calculated spectrum in the region 50–250 keV

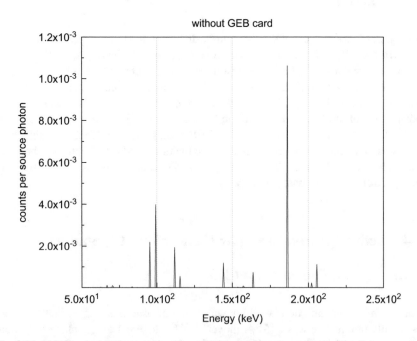

Fig. 3.16 MCNP calculated spectrum without GEB special treatment for the F8 tally

Example 3.4 Cf Shuffler Input

```
vr03.txt for 500 g 235U, density 2.5g/cm^3
c
c       Cells
c
10     5 -2.5   -10        imp:n=1  $ SNM
11     0        -12 10     imp:n=1  $ void about SNM
12     2 -8.65 -11 12      imp:n=1  $ Cd
13     1 -0.94 -13 11 100 101 102 103 104 105 106 107 108
       109 110 111   imp:n=1  $ poly
c
c      3He Tubes
c
100    3 1.002e-4 -100     imp:n=1 $ #1
101    3 1.002e-4 -101     imp:n=1 $ #2
102    3 1.002e-4 -102     imp:n=1 $ #3
103    3 1.002e-4 -103     imp:n=1 $ #4
104    3 1.002e-4 -104     imp:n=1 $ #5
105    3 1.002e-4 -105     imp:n=1 $ #6
106    3 1.002e-4 -106     imp:n=1 $ #7
107    3 1.002e-4 -107     imp:n=1 $ #8
108    3 1.002e-4 -108     imp:n=1 $ #9
109    3 1.002e-4 -109     imp:n=1 $ #10
110    3 1.002e-4 -110     imp:n=1 $ #11
111    3 1.002e-4 -111     imp:n=1 $ #12
c
9999   0 13       imp:n=0   $ outside cell
c


c
c      Surfaces
c
10 rcc 0 0 7      0 0 7.073553    3      $ SNM
11 rcc 0 0 6      0 0 9           4      $ exterior of Cd
12 rcc 0 0 6.1    0 0 8.8         3.9    $ interior of Cd
13 rcc 0 0 0      0 0 21          14  $ exterior of diagnostic
c
c      3He Tubes
c
100 rcc 8.270    0.000 0.100  0 0 20.8    1.27   $ #1
101 rcc 7.162    4.135 0.100  0 0 20.8    1.27   $ #2
102 rcc 4.135    7.162 0.100  0 0 20.8    1.27   $ #3
103 rcc 0.000    8.270 0.100  0 0 20.8    1.27   $ #4
```

```
104 rcc -4.135   7.162 0.100   0 0 20.8    1.27    $ #5
105 rcc -7.162   4.135 0.100   0 0 20.8    1.27    $ #6
106 rcc -8.270   0.000 0.100   0 0 20.8    1.27    $ #7
107 rcc -7.162  -4.135 0.100   0 0 20.8    1.27    $ #8
108 rcc -4.135  -7.162 0.100   0 0 20.8    1.27    $ #9
109 rcc 0.000   -8.270 0.100   0 0 20.8    1.27    $ #10
110 rcc 4.135   -7.162 0.100   0 0 20.8    1.27    $ #11
111 rcc 7.162   -4.135 0.100   0 0 20.8    1.27    $ #12
c


c
c       Materials
c
m1      6000.  1   1001. 2 $ Polyethylene
mt1     poly.10t                $ S(a,b)
c
m2      48106.  0.0125
        48108.  0.0089
        48110.  0.1249
        48111.  0.128
        48112.  0.2413
        48113.  0.1222
        48114.  0.2873
        48116.  0.0749
c
m3      2003.  1               $ He-3
c
m5      92235. 1.
c
mode    n
prdmp   j j j 4 j
print
nps 1e10
c
c       -------------------------------------------------
c   Cf source
phys:n    30 30 0 J J J 0 30 J J J 0 0
totnu
sdef pos=d1 tme=d2 erg=d3
sp3     -3 1.175 1.0401  $ Frohner Watt parameters
si1  L  3.8 0 11
sp1     1.0
si2   0e8  10e8  $ upper bin boundaries (no option = option H)
sp2   0    1     $ bin probabilities (no option = option D, H, L)
```

```
c       ------------------------------------------------------------------
c
c       Tallies
fc4     <<<<< Detection Rate per source Cf neutron >>>>>
f4:n    (100 101 102 103 104 105 106 107 108 109 110 111)
fm4     -1 3 103
sd4     1
t4      10e8 98i 1000e8
tf4     7j 7
fq4     t f
c
```

The geometry is very simple using an (unrealistic) ^{235}U sample of density 2.5 g/cm³ surrounded by 12 simplified ^3He tubes embedded in polyethylene, as shown in Fig. 3.17. The detection chamber is lined with cadmium.

The californium irradiation position is in the cavity but outside of the item, at the mid-height. There is nothing particular to this case in the model geometry or cross sections. One parameter in the input file is the use of the TOTNU card (MCNP6 default option), which generates both prompt and delayed neutrons from fission. Delayed neutron options are controlled by the ACT card:

```
ACT DN = [MODEL|LIBRARY|BOTH|PROMPT]
```

If MODEL is used, production of delayed neutrons uses models only. If LIBRARY is used, production of delayed neutrons uses library only (default option). If BOTH is used, production of delayed neutrons uses models when libraries are missing. If PROMPT is used, both prompt and delayed neutrons are treated as prompt. In this example, we use the default option.

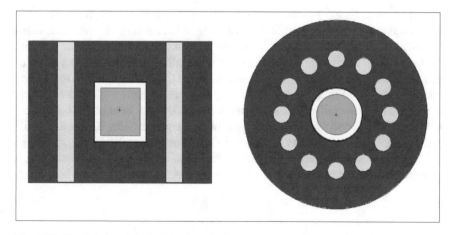

Fig. 3.17 Simplified model of californium shuffler

The neutron source information is described on the following cards:

```
sdef pos=d1 tme=d2 erg=d3
sp3     -3 1.175 1.0401 $ Frohner Watt parameters
si1   L  3.8 0 11
sp1      1.0
si2      0e8 10e8   $ upper bin boundaries (no option = option H)
sp2      0    1     $ bin probabilities (no option = option D, H, L)
```

The neutron energy spectrum of Cf is described by the d3 distribution, which is a Watt distribution with parameters from Frohner (Ref. [4] of Chap. 2) (*NOTE: Some versions of the MCNP(X) manuals gave incorrect Watt parameters for the Cf spectrum, resulting in the wrong average energy*). An alternative source specification could have been to use PAR = SF (with Cf in the material of the source location) in order to obtain the built-in energy (and multiplicity) information for Cf. In this calculation, we do not include the practical effect of bringing the source to the irradiation position or removing it. The irradiation starts instantly at time zero and ends completely after 10 s (10E8 shakes).

The tally cards are shown here:

```
c       Tallies
fc4     <<<<< Detection Rate per source Cf neutron >>>>>
f4:n    (100 101 102 103 104 105 106 107 108 109 110 111)
fm4     -1 3 103
sd4      1
t4       10e8 98i 1000e8
tf4      7j 7
fq4     t f
```

The f4 track length estimator tally is used with the fm4 card to calculate the ^{3}He(n,p) rate in the sum of all ^{3}He detectors. The volume is set to 1.0 on the sd4 card to give the total reaction (rather than the rate per unit volume). The time bins on this card are chosen to give the ^{3}He rates during the irradiation period and in a series of counting intervals after the "removal" of the source. The tf4 card is used to produce tally fluctuation information for a counting interval of interest (the seventh time bin). By basing the tally fluctuation charts on the seventh time bin, all statistical sampling information and statistical tests of convergence (Print Tables 160, 161, and 162) will be based on the seventh time bin. This bin is tally in the time range 60–70 s. Otherwise, the tally fluctuation charts and all statistical information and convergence information would be based on the default last (total) time bin, which is less relevant to the results in this problem. The fq4 card is used to make the time bins print out in a vertical column, which is much easier to cut and paste into other applications.

3.4.2   Results

The problem summary table is shown in Fig. 3.18. We can see that the weight of prompt fission neutrons created was 7.8E−02 and the weight of delayed neutrons was 5.02E−04. This result gives a delayed fraction of 0.0064, which agrees with the expected value for fast fission (0.0064) in ^{235}U (see, e.g., [6]). We would expect fast fission to dominate because the cavity is lined with cadmium. (The beta value for thermal fission is very similar at 0.0065.) This is a check that the selected cross-section data files contain the necessary data.

The results of the F4 tally are given in Fig. 3.19. The reaction rate in the first time bin (while the source is present) is large and has a very small statistical uncertainty, which dominates the total over all time bins. The values in the later time bins are much smaller and the uncertainties are much larger, which is why it is important to use the tf4 card so that the statistical tests are shown for a quantity relevant to the actual result. The results of the tally fluctuation analysis are given in Fig. 3.20.

To obtain reasonable statistical quality results on the later time bins, a very large number of starting particles have been used (1E10), which is a brute force way of obtaining the results. The later tally bins have large errors but make very small contribution to the final result. This problem is probably a suitable candidate for variance reduction techniques to use computer time more effectively.

The tally bins give the number of counts recorded in each time window per source neutron for one irradiation cycle. We can use these values to calculate the counts in a real measurement with multiple irradiation cycles (see Fig. 3.21).

The counts recorded after the first irradiation are in the 10–20 s time bin. The counts after the second irradiation are the sum of the counts in 10–20 s time bin (from irradiation #2) plus the 30–40 s time bin, which represents the contribution to the second counting time from the first irradiation. The counts after the third irradiation are the sum of the 10–20 s bin (from third irradiation) plus the 30–40 s bin (from the second irradiation) plus the 60–70 s bin (from the first irradiation) and so on. With a Cf source strength of 10^8 neutrons/s, the number of counts recorded in each counting interval is shown in Table 3.5 and plotted in Fig. 3.22. We can see that the counts in each interval increase asymptotically as the delayed neutrons groups with longer half-lives build up. The experimental result would be equivalent to the sum of the counts up to the actual number of irradiation cycles used.

neutron creation	tracks	weight (per source particle)	energy (per source particle)
source	10000000000	1.0000E+00	2.1215E+00
nucl. interaction	0	0.	0.
particle decay	0	0.	0.
weight window	0	0.	0.
cell importance	0	0.	0.
weight cutoff	0	0.	0.
e or t importance	0	0.	0.
dxtran	0	0.	0.
forced collisions	0	0.	0.
exp. transform	0	0.	0.
upscattering	0	0.	2.2859E-07
photonuclear	0	0.	0.
(n,xn)	3301290	3.3013E-04	2.5309E-04
prompt fission	782550962	7.8255E-02	1.5968E-01
delayed fission	5018271	5.0183E-04	2.5553E-04
prompt photofis	0	0.	0.
tabular boundary	0	0.	0.
tabular sampling	0	0.	0.
total	10790870523	1.0791E+00	2.2817E+00

neutron loss	tracks	weight (per source particle)	energy (per source particle)
escape	4969414442	4.9694E-01	5.4988E-01
energy cutoff	0	0.	0.
time cutoff	0	0.	0.
weight window	0	0.	0.
cell importance	0	0.	0.
weight cutoff	0	0.	0.
e or t importance	0	0.	0.
dxtran	0	0.	0.
forced collisions	0	0.	0.
exp. transform	0	0.	0.
downscattering	0	0.	1.7019E+00
capture	5507096339	5.5071E-01	5.6403E-03
loss to (n,xn)	1649823	1.6498E-04	1.3942E-03
loss to fission	317709919	3.1271E-02	2.2902E-02
nucl. interaction	0	0.	0.
particle decay	0	0.	0.
tabular boundary	0	0.	0.
elastic scatter	0	0.	0.
total	10790870523	1.0791E+00	2.2817E+00

number of neutrons banked	476510781
neutron tracks per source particle	1.0791E+00
neutron collisions per source particle	5.3622E+01
total neutron collisions	536217496592
net multiplication	1.0477E+00 0.0000

average time of (shakes)

escape	5.0035E+08
capture	5.0078E+08
capture or escape	5.0058E+08
any termination	4.8600E+08

cutoffs

tco	1.0000E+33
eco	0.0000E+00
wc1	-5.0000E-01
wc2	-2.5000E-01

Fig. 3.18 Problem summary table

```
ltally        4          nps • 10000000000
    <<<<< Detection Rate per source Cf neutron >>>>>
    tally type 4    track length estimate of particle flux.
    particle(s): neutrons
    cell a is (100 101 102 103 104 105 106 107 108 109 110 111)

            volumes
                  cell:        a
                       1.00000E+00

multiplier bin:  -1.00000E+00        3         103
      cell:            a
      time
    1.0000E+09   2.18723E-01   0.0000
    2.0000E+09   3.80251E-05   0.0020
    3.0000E+09   8.91106E-06   0.0042
    4.0000E+09   4.70598E-06   0.0058
    5.0000E+09   3.03004E-06   0.0072
    6.0000E+09   2.10977E-06   0.0087
    7.0000E+09   1.57281E-06   0.0101
    8.0000E+09   1.16845E-06   0.0117
    9.0000E+09   8.73593E-07   0.0136
    1.0000E+10   6.73758E-07   0.0154
    1.1000E+10   5.12576E-07   0.0178
    1.2000E+10   3.93958E-07   0.0200
    1.3000E+10   3.13749E-07   0.0223
    1.4000E+10   2.38708E-07   0.0254
    1.5000E+10   2.01865E-07   0.0285
    1.6000E+10   1.57759E-07   0.0318
    1.7000E+10   1.31750E-07   0.0352
    1.8000E+10   1.07695E-07   0.0377
    1.9000E+10   8.48956E-08   0.0435
    2.0000E+10   7.77946E-08   0.0473
    2.1000E+10   5.50909E-08   0.0510
    2.2000E+10   4.83979E-08   0.0548
    2.3000E+10   4.07598E-08   0.0620
    2.4000E+10   3.85655E-08   0.0668
    2.5000E+10   3.26456E-08   0.0692
    2.6000E+10   2.60775E-08   0.0858
    2.7000E+10   2.34774E-08   0.0937
    2.8000E+10   2.06355E-08   0.0857
    2.9000E+10   2.02929E-08   0.0884
    3.0000E+10   1.39312E-08   0.0923
```

Fig. 3.19 Results of Tally 4 in shuffler problem (shown up to 300 s after the irradiation)

1status of the statistical checks used to form confidence intervals for the mean for each tally bin

tally result of statistical checks for the tic bin (the first check not passed is listed) and error magnitude check for all bins

    4 passed the 10 statistical checks for the tally fluctuation chart bin result
    passed all bin error check: 101 tally bins had   22 bins with zeros and 48 bins with relative errors exceeding 0.10

the 10 statistical checks are only for the tally fluctuation chart bin and do not apply to other tally bins.

the tally bins with zeros may or may not be correct: compare the source, cutoffs, multipliers, et cetera with the tally bins.

warning.   1 of the   1 tallies had bins with relative errors greater than recommended.
1tally fluctuation charts

```
                  tally        4
    nps       mean       error    vov    slope   fom
1000000000  1.6295E-06  0.0311  0.0028   10.0   2.8E-01
2000000000  1.6365E-06  0.0220  0.0014   10.0   2.8E-01
3000000000  1.6213E-06  0.0181  0.0010   10.0   2.7E-01
4000000000  1.5920E-06  0.0159  0.0008   10.0   2.7E-01
5000000000  1.5920E-06  0.0142  0.0006   10.0   2.7E-01
6000000000  1.5835E-06  0.0130  0.0005   10.0   2.7E-01
7000000000  1.5851E-06  0.0120  0.0004   10.0   2.7E-01
8000000000  1.5777E-06  0.0112  0.0004   10.0   2.7E-01
9000000000  1.5763E-06  0.0106  0.0003   10.0   2.7E-01
10000000000 1.5728E-06  0.0101  0.0003   10.0   2.7E-01
```

Fig. 3.20 Tally fluctuation chart (seventh time bin)

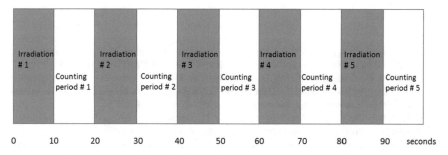

Fig. 3.21 Multiple irradiation and counting cycles

Table 3.5 Counts recorded for each irradiation cycle

Irradiation #	Counts	Irradiation #	Counts	Irradiation #	Counts	Irradiation #	Counts
1	3803	8	4747	15	4781	22	4785
2	4273	9	4758	16	4782	23	4785
3	4484	10	4766	17	4783	24	4785
4	4601	11	4771	18	4783	25	4785
5	4668	12	4775	19	4784	26	4785
6	4708	13	4777	20	4784	27	4785
7	4732	14	4779	21	4784	28	4785

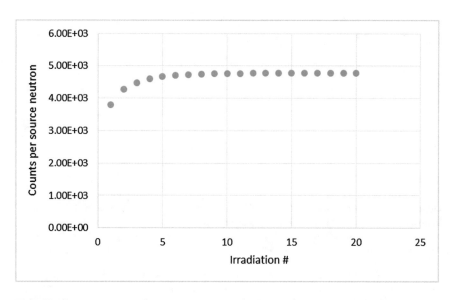

Fig. 3.22 Counts in each cycle

References

1. P. Santi, D. Beddingford, D. Mayo, *Revised Prompt Neutron Emission Multiplicity Distributions for 236, 238-Pu*, Los Alamos National Laboratory Report LA-UR-04-8040 (2004)
2. D. Reilly, N. Ensslin, H. Smith Jr., S. Kreiner, *Passive Nondestructive Assay of Nuclear Materials*, LA-UR-90-732 (U.S. Government Printing Office, Washington, DC, 1991) https://www.lanl.gov/org/ddste/aldgs/sst-training/technical-references.php
3. K. Boehnel, The Effect of Multiplication on the Quantitative Determination of Spontaneoulsy Fissioning Isotopes by Neutron Correlation Analysis. Nucl. Sci. Eng. **90**, 75–82 (1985)
4. M. Krick, W. Harker, W. Geist, J. Longo, *International Neutron Coincidence Counting: Software Users Manual*, Los Alamos Laboratory Report LA-UR-01-6761 (2009)
5. P. Rinard, Chapter 9: Shufflers, in *Passive Nondestructive Assay of Nuclear Materials 2007 Addendum*, LA-UR-03-4404, ed. by D. Reilly, (2007) https://www.lanl.gov/org/ddste/aldgs/sst-training/_assets/docs/PANDA%202007%20Addendum/0.%20PANDA%20Preface%20v1-3.pdf
6. J.R. Lamarsh, Introduction to Nuclear Reactor Theory. Addison-Wesley Series, in *Nuclear Engineering*, (Addison-Wesley Pub, Reading, MA, 1966)

Chapter 4
Examples of Advanced Concepts

4.1 Variance Reduction

4.1.1 Introduction

Variance reduction techniques are methods that reduce the variance (or statistical relative error) for a given amount of computational time in a Monte Carlo calculation. This is equivalent to reducing the computational time required to achieve a given statistical error. Variance-reduction methods work by sampling more often the important parts of a problem at the expense of less important parts.

There are four classes of variance reduction techniques:

- *Truncation:* The calculation omits parts of the problem that do not contribute significantly to the desired result. Geometry is always truncated. Energy may be truncated by eliminating—or killing—all histories whose energy falls below a specified limit, which is very dangerous in neutron problems with fission. For photon problems, all photoatomic reactions less than 1 keV are generally considered to be absorption and cut off. Time may be cut off to not follow tracks after they are no longer tallied. Physics is generally cut off by ignoring photonuclear effects or secondary particle production, etc.
- *Population control:* The number of statistical samples in any part of the problem may be reduced if the weight of the samples is increased to preserve total weight and yield the correct expected value of the tallies. The number of samples may also be increased if the weight of the samples is reduced to preserve total weight and yield the expected value of the tallies. Consequently, there can be many samples of low weight in important parts of a problem and less in unimportant parts. Population control methods include importance sampling, weight windows, time and energy splitting, secondary particle production, etc.

Supplementary Information The online version contains supplementary material available at [https://doi.org/10.1007/978-3-031-04129-7_4].

J. S. Hendricks et al., *Monte Carlo N-Particle Simulations for Nuclear Detection and Safeguards*, https://doi.org/10.1007/978-3-031-04129-7_4

- *Modified sampling:* The statistical sampling algorithms may be modified to favor different parts of a problem. Source directional biasing sends more of the source in a desired direction and less in an undesired direction. Other source biasing modifies the sampling algorithm to produce more of the source in desired energy, space, time, or other problem phase space regions. The exponential transform can stretch the path of neutral particles in the desired direction and shrink it in undesired directions. Sampling of secondary particle production can alter the energy or particle type, etc.
- *Partially deterministic methods:* In special cases, an estimate of the particle tracks that would reach a specified location can be used to directly place them there. Next-event estimators put estimates at a point or particles on a surface after a source or collision event, weighted with the probability that they would arrive there without further interaction.

Problems where all phase space is equally important do not need variance reduction. Examples are reactor criticality, problems with highly efficient detectors, or very simple problems. Problems with pulse-height tallies, coincidence detectors, or other physics that relies on coincidences cannot use variance reduction, but for problems that need better resolution or faster computation and where parts of the problem contribute more than others, variance reduction is usually essential.

Applying variance reduction can be challenging. It is difficult to know what parts of the problem are important unless the problem has already been solved, in which case, further calculations may not be needed. A poor choice of variance reduction techniques can lead to less efficient or even falsely converged results. Good intuition and iteration are often needed. Fortunately, the MCNP code has many diagnostics and outputs to provide insight into problems [1]. In many cases—particularly radiation shielding problems or when sources and detectors are far away or out of sight—variance reduction can speed up problems by many orders of magnitude and enable much higher resolution.

The MCNP code can generate variance-reduction parameters with the weight window generator. Weight windows are a population control method in which a range of desired weights is specified for each region of space, time, and/or energy in a problem. Particle tracks that enter a region where the particle weight is higher than the weight window are split into more particles with lower weights now closer to or in the window. Particles that enter with weight lower than the window undergo Russian roulette so that some survive with higher weights, again closer to or in the window. The weight window generator works by keeping track of all the particle weight that passes through a region of space, time, or energy and also keeping track of the particle weight that passes through that eventually leads to a score in a target tally.

The weight window bound is then set by

$$\frac{\left(\text{weight passing through}\right)}{\left(\text{weight passing through that scores}\right)}.$$

These bounds are then renormalized so that source particles are born in the weight window and split as they approach the tally, yielding many low-weight samples at the tally.

In the lead slowing-down spectrometer (LSDS) example (see Sect. 4.1.2), variance reduction sped the calculation by a factor of more than 1000—a calculation that would take a day could now be run in a minute. An energy bin could be divided into 30 smaller bins with no loss of resolution. For some problems, variance reduction is the only way to achieve convergence.

Although applying variance reduction may not appear straightforward, measuring its effectiveness is. The only measure of Monte Carlo efficiency is the FOM:

$$FOM = 1 / \left(E^2 T \right)$$

where E is the statistical error and T is the computational time. For a problem with N statistical samples, T is proportional to N and E^2 is proportional to $1/N$. Consequently, the FOM converges to a constant dependent on the problem efficiency and the computer efficiency. Efficient problems have low T and low E and consequently a high FOM.

The goal of variance reduction is to maximize the FOM. In so doing, it is important to keep the number of tracks constant from the source region to the tally region.

- The number of tracks should decrease and particle weights should increase in unimportant regions of space, energy, and time.
- The number of tracks should increase and particle weights should decrease in important regions.
- The weights between adjacent regions should be no more than a factor of 4 different.

It is important to achieve adequate sampling of all cells and other parts of phase space and to ensure that tallies have passed the statistical checks or be able to justify why not.

Variance reduction is an art as well as a science and there are many possible approaches as can be found searching the internet for "variance reduction MCNP." A useful resource on this topic is the reference collection of documents at mcnp. lanl.gov. An external tool is ADVANTG [2]. We recommend the following strategy to variance reduction:

- Simplify the problem if possible and use a much simpler tally to get an answer with 10–20% statistical errors in a few minutes. Optimize the problem with a tally closer in phase space to the source and with less resolution. Consider using the weight window generator to learn what is important in the problem; often the generated weight windows will provide the needed variance reduction technique as well.
- Once the modified problem gives 10–20% relative error on the optimization tally, carefully examine the output, and then move or otherwise make the tally more like the final one in the problem. Add only one variance reduction technique at a time unless more are clearly needed.
- After a few short iterations of 1–5 min each, the full problem will now be optimized and may be significantly more efficient.

4.1.2   *Multigroup Weight Windows and Time Splitting: Lead Slowing-Down Spectrometer*

Example 4.1 is a Lead Slowing-Down Spectrometer (LSDS) analyzing a spent fuel assembly. Variance reduction is the only way to achieve acceptable convergence. In this case, multigroup weight windows and time splitting are applied. The steps taken to choose a good variance-reduction approach are presented.

Example 4.1 Input for Lead Slowing-Down Spectrometer (No Variance Reduction)

```
Lead Slowing Down Spectrometer LSDS
c with PWR 17x17 fuel element
c *********** fuel, u=1
10 200 -11.00 -1 60 -61 u=1  imp:n=1 vol=4.17441E+03 $ fuel
c                                              (middle section)
11 200 -11.00   -1 -60   u=1  imp:n=1 $ fuel (bottom end)
12 200 -11.00   -1  61   u=1  imp:n=1 $ fuel (top end)
14 300 -6.44    +1 -3    u=1  imp:n=1
16   0          +3       u=1  imp:n=1 $ outside fuel pin
c *********** guide tube, u=8
20   0          -5       u=8  imp:n=1 $ inside guide tube
22 300 -6.44   +5 -6     u=8  imp:n=1 $ guide tube
26   0          +6       u=8  imp:n=1 $ outside guide tube
c *********** lattice, u=10, fill with u=1,8
100  0          -50      lat=1 u=10   $ lattice of fuel + guide
     fill = -8:8 -8:8 0:0
     1 1 1 1 1 1 1 1 1 1 1 1 1 1 1 1 1
     1 1 1 1 1 1 1 1 1 1 1 1 1 1 1 1 1
     1 1 1 1 1 8 1 1 8 1 1 8 1 1 1 1 1
     1 1 8 1 1 1 1 1 1 1 1 1 8 1 1 1
     1 1 1 1 1 1 1 1 1 1 1 1 1 1 1 1 1
     1 1 8 1 1 8 1 1 8 1 1 8 1 1 8 1 1
     1 1 1 1 1 1 1 1 1 1 1 1 1 1 1 1 1
     1 1 1 1 1 1 1 1 1 1 1 1 1 1 1 1 1
     1 1 8 1 1 8 1 1 8 1 1 8 1 1 8 1 1
     1 1 1 1 1 1 1 1 1 1 1 1 1 1 1 1 1
     1 1 1 1 1 1 1 1 1 1 1 1 1 1 1 1 1
     1 1 8 1 1 8 1 1 8 1 1 8 1 1 8 1 1
     1 1 1 1 1 1 1 1 1 1 1 1 1 1 1 1 1
     1 1 1 8 1 1 1 1 1 1 1 1 8 1 1 1
     1 1 1 1 1 8 1 1 8 1 1 8 1 1 1 1 1
     1 1 1 1 1 1 1 1 1 1 1 1 1 1 1 1 1
     1 1 1 1 1 1 1 1 1 1 1 1 1 1 1 1 1
     imp:n=1
```

```
102  0                -55   fill=10 imp:n=1
c **Fission chamber detectors - place in cartesian coord
199  0                -110           u=2 imp:n=1 $ interior
200  125 -2.7         +110           u=2 imp:n=1 $ detector casing
201  0                -100    fill=2 imp:n=1 $ single detector universe=
202 like 201 but trcl=( 4.2686 -0.4204 0) imp:n=1
203 like 201 but trcl=( 8.3731 -1.6655 0) imp:n=1
204 like 201 but trcl=( 12.1559 -3.6874 0) imp:n=1
205 like 201 but trcl=( 15.4715 -6.4085 0) imp:n=1
206 like 201 but trcl=( 18.1926 -9.7241 0) imp:n=1
207 like 201 but trcl=( 20.2145 -13.5069 0) imp:n=1
208 like 201 but trcl=( 21.4596 -17.6114 0) imp:n=1
209 like 201 but trcl=( 21.8800 -21.8800 0) imp:n=1
210 like 201 but trcl=( 21.4596 -26.1486 0) imp:n=1
211 like 201 but trcl=( 20.2145 -30.2531 0) imp:n=1
212 like 201 but trcl=( 18.1926 -34.0359 0) imp:n=1
213 like 201 but trcl=( 15.4715 -37.3515 0) imp:n=1
214 like 201 but trcl=( 12.1559 -40.0726 0) imp:n=1
215 like 201 but trcl=( 8.3731 -42.0945 0) imp:n=1
216 like 201 but trcl=( 4.2686 -43.3396 0) imp:n=1
217 like 201 but trcl=( 0.0000 -43.7600 0) imp:n=1
218 like 201 but trcl=( -4.2686 -43.3396 0) imp:n=1
219 like 201 but trcl=( -8.3731 -42.0945 0) imp:n=1
220 like 201 but trcl=( -12.1559 -40.0726 0) imp:n=1
221 like 201 but trcl=( -15.4715 -37.3515 0) imp:n=1
222 like 201 but trcl=( -18.1926 -34.0359 0) imp:n=1
223 like 201 but trcl=( -20.2145 -30.2531 0) imp:n=1
224 like 201 but trcl=( -21.4596 -26.1486 0) imp:n=1
225 like 201 but trcl=( -21.8800 -21.8800 0) imp:n=1
226 like 201 but trcl=( -21.4596 -17.6114 0) imp:n=1
227 like 201 but trcl=( -20.2145 -13.5069 0) imp:n=1
228 like 201 but trcl=( -18.1926 -9.7241 0) imp:n=1
229 like 201 but trcl=( -15.4715 -6.4085 0) imp:n=1
230 like 201 but trcl=( -12.1559 -3.6874 0) imp:n=1
231 like 201 but trcl=( -8.3731 -1.6655 0) imp:n=1
232 like 201 but trcl=( -4.2686 -0.4204 0) imp:n=1
c *********** fission chamber detector casing
300 400 -8.03 -200 +201 -202 +203  imp:n=1 $ outer SS casing
305 405 -8.96 -201 -202 +212        imp:n=1 $ Cu disk
310 410 -4.54 -201 -212 +222        imp:n=1 $ TiD_2 disk
315 220 -0.001205 -201 +203 -222    imp:n=1 $ air in SS casing
500  0 55 -500 vol=2.61853E+04      imp:n=1 $ inside LSDS
c *********** surrounding Pb
```

```
502   100 -11.34 +500 -520
      #201 #202 #203 #204 #205 #206 #207 #208 #209 #210
      #211 #212 #213 #214 #215 #216 #217 #218 #219 #220
      #221 #222 #223 #224 #225 #226 #227 #228 #229 #230
      #231 #232                                      imp:n=1
504   100 -11.34 +500 +520 -540                      imp:n=1
506   100 -11.34 +500 +540 -560 (+200:+202:-203)     imp:n=1
999   0             +560  (+200:+202:-203)           imp:n=0


1 rcc 0 0 -50     0 0 100    0.4096   $ fuel
2 rcc 0 0 -50     0 0 100    0.4178   $ gap
3 rcc 0 0 -50     0 0 100    0.4750   $ cladding
5 rcc 0 0 -50     0 0 100    0.5715   $ guide inner
6 rcc 0 0 -50     0 0 100    0.6120   $ guide outer
10 rcc 0 0 -50    0 0 100    0.6120   $ instrum tube
50 rpp  -0.6299   0.6299  -0.6299  0.6299 -50 50 $ one pin
55 rpp -10.7083 10.7083 -10.7083 10.7083 -50 50 $ assembly
60 pz -15                            $ fuel segment lower-mid
61 pz  15                            $ fuel segment mid-upper
100 rcc 0 21.88 -30   0 0 60   1.945 $ assay detector outside
110 rcc 0 21.88 -30   0 0 60   1.645 $ assay detector inside
c *********** DT Neutron source
200 cx 4.0000      $ outside of SS casing
201 cx 3.6825      $ inside of SS casing
202 px -70.0000    $ edge of casing near fuel bore
203 px -100.000    $ edge of casing at end of LSDS
212 px -70.6350    $ Cu disk
222 px -70.6354    $ TiD_2 disk
c *********** LSDS8
500 rcc 0 0 -50   0 0 100    15.1443 $ inside of LSDS
520 rcc 0 0 -50   0 0 100    25.0    $ segment 1 of LSDS
540 rcc 0 0 -50   0 0 100    45.0    $ segment 2 of LSDS
560 rcc 0 0 -50   0 0 100    100.0   $ outside LSDS


m0 nlib=.70c
c  Air, rho=0.001205
m220 7014 -0.755267 8016 -0.231781 18040 -0.012827
m100 82208 .524 82207 .221 82206 .241 82204 .014     $ Pb
m125 13027 1                             $ Aluminum rho=2.7 g/cm3
m200 92238 -4.360E+05   8016 -6.186E+04 92235 -3.248E+03
     92236 -1.790E+03 94239 -2.336E+03 94240 -1.092E+03
     54136 -1.092E+03 42000.66c -1.197E+03
c   Zirc4
m300 40090 .5145 40091 .1122 40092 .1715 40094 .1738 40096 .0280
```

```
c Neutron tube materials: stainless steel, type 304 rho=8.03 g/cm3
m400 26056 .678950 26054 .043290 26057 .015688 26058 .002072 $ Fe
     24052 .150820 24053 .017102 24050 .007821 24054 .004259 $ Cr
     28058 .054462 28060 .020978 28062 .002908 28061 .000912 $ Ni
     28064 .000740 $ stainless steel tube, type 304 rho=8.03
m405 29063 .6915 29065 .3085 $ Copper, rho=8.96 g/cm3
c      Ti_D2:
m410 1002 2 22048 .74 22046 .08 22047 .07 22049 .05 22050 .05
m5 92235 1 $ used for tallying fission rates and neutron production
c
sdef pos=D1 erg FPOS = D2 tme=D6
si1 L -70.6352 0 0   -70.6352 0 0   -70.6352 0 0
sp1    4         2         100
DS2 S 3  4   5
si3 H 0 1 2 3 4 5 6 7 8 9
sp3 D 0 .445747263 .648944683 .840905395 .969855575 .9956059
        .909679323 .739792038 .535490104 .344995724
sp4 -4 0.21 2.45
sp5 -4 0.45 14.0
si6 H 0 100
sp6 D 0 1
c
c    default tally bins
t0   1e2 263ilog 2e5
e0   1e-4 1e-3 1e-2 1e2
fq0 T E
c
fc14 Average flux in fuel
f14:n 10
c
fc44 Average flux in fission chambers
f44:n (201  202  203  204  205  206  207  208  209  210
       211  212  213  214  215  216  217  218  219  220
       221  222  223  224  225  226  227  228  229  230
       231  232)
c
fc114 U235 fission rate
f114:n (201  202  203  204  205  206  207  208  209  210
        211  212  213  214  215  216  217  218  219  220
        221  222  223  224  225  226  227  228  229  230
        231  232)
fm114 1 5 (-6)
c
```

```
fc214 U235 fission nubar
f214:n (201   202   203   204   205   206   207   208   209   210
         211   212   213   214   215   216   217   218   219   220
         221   222   223   224   225   226   227   228   229   230
         231   232)
fm214 1 5 (-6 -7)
c
cut:n 200000
prdmp 2j 1 3
print -128 -162 -30
nps 1e8
```

The geometry of Example 4.1 is shown in cut-away view in Fig. 4.1 and in axial view in Fig. 4.2. Axial views of each repeated structures level are shown in Figs. 4.3, 4.4, and 4.5.

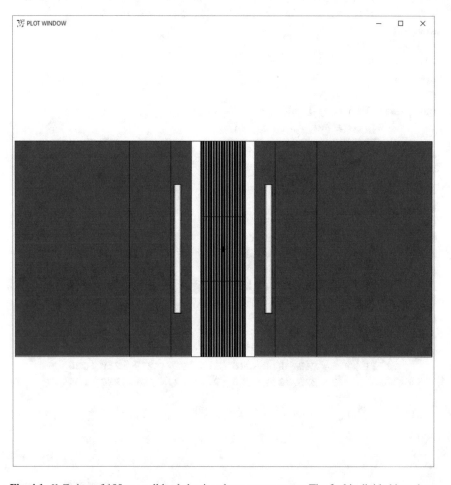

Fig. 4.1 Y–Z view of 100 cm–tall lead slowing-down spectrometer. The fuel is divided into three Z-axis sections for later depletion calculations, and the lead is divided into three radial sections for variance reduction by importance sampling (not used)

4.1.2.1 Input File Notes

The LSDS geometry file consists of 53 cells 22 surfaces, which the MCNP code expands to 192 surfaces after LIKE BUT duplication and macrobody expansion into quadratic surfaces. The cells and surfaces at the three different geometry levels are illustrated in Figs. 4.1, 4.2, 4.3, 4.4, and 4.5.

Note how the volume is specified in the cell descriptions for cell 10 and cell 500. If it is not specified, the MCNP code calculates the volume. The MCNP code could not calculate the volume of cell 500—the annulus containing the fission chambers—because that cell is neither rotationally symmetric nor a polyhedron. This volume is not used for anything other than calculating cell mass. The volume of the fuel element cell 10 is input as 4174.41 cm^3, whereas the MCNP code calculates it to be 15.8122 cm^3. This difference is because the MCNP code calculates the volume

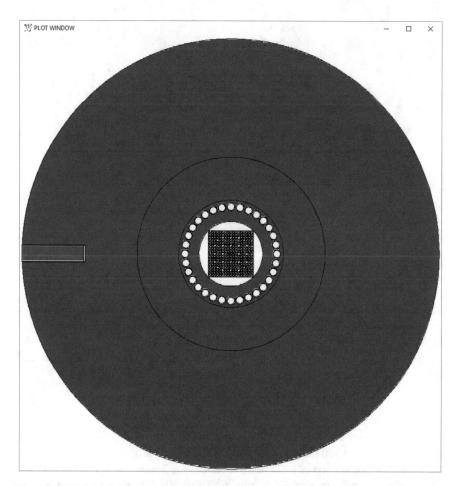

Fig. 4.2 *X–Y* view of 100 cm radius lead slowing-down spectrometer. The lead (blue) is divided into three radial sections for variance reduction by importance sampling (not used). The ring of 32 fission detectors surrounds the fuel cavity with a 17 × 17 pressurized-water reactor spent fuel assembly inside. The fission detectors are modeled with void surrounded by a thin aluminum casing. The red section is space for the neutron generator (actual source location in yellow)

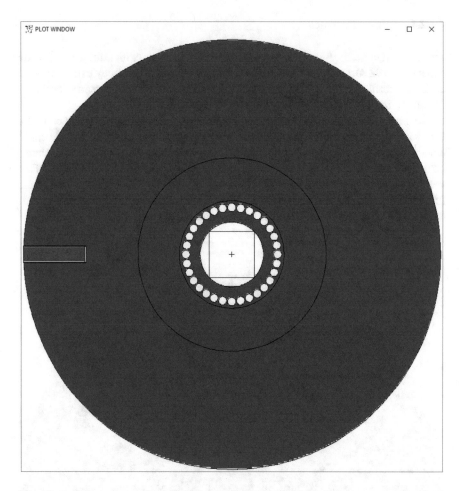

Fig. 4.3 *X–Y* view universe level *u* = 0 consisting of the lead slowing-down spectrometer, cylindrical holes for the fission detectors, and a square RPP box for the fuel assembly

of a single fuel pin, but the MCNP code needs the volume of all 264 fuel pins to properly calculate the flux in the fuel, tally F14, so the volume is input 264 times higher. Alternatively, the volume could be specified on as "SD14 4.17441E3" with the SD14 tally card for tally 14. Or the volume of the first (cell 10) and 49th (cell 500) cells could be set on a VOL card in the data portion of the input file:

```
vol 4.17441E+03 47j 2.61853E+04 4j
```

The materials use ENDF/B-VII neutron cross sections, which is set by NLIB = .70c on the M0 material card. The M0 material card sets the default for all nuclides. If a .70c version of a nuclide does not exist, then its full specification must be provided, as in the case of 42000.66c. ENDF/B-VII data are mostly isotopic, so

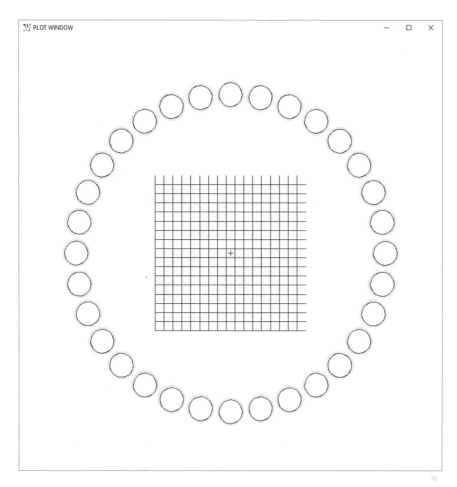

Fig. 4.4 *X–Y* view universe level 1 consisting of universes *u* = 2 and *u* = 10. The plot extent is 60 × 60 (EX = 30) rather than 200 × 200 (EX = 100) as in Figs. 4.1, 4.2, and 4.3. The fission chambers, universe *u* = 2, have an aluminum lining (yellow). The lattice for the fuel assemblies is *u* = 10

iron, chromium, nickel, etc., must be expressed in their isotopic constituents and fractions. Whereas molybdenum is <1% of material m200, it is specified as an element, which requires an earlier 2001 ENDF/B-VI evaluation, 42000.66c. The original LSDS model employed for safeguards research had a spent fuel assembly with the depletion performed on the fuel, material m200, and the zircaloy clad, material m300. The fuel had 158 isotopes, mostly fission products and actinides; the clad had 108 isotopes, many in trace quantities <1 × 10^{-10} g/cm^3. A real safeguards study would run burnup/depletion calculations to get the clad and spent fuel isotopes at three elevations for each fuel pin, which would each be its own universe so that the depleted fuel could differ pin to pin and vertically. The total material description would be tens of thousands of lines long. For convenience, only a simplified fuel

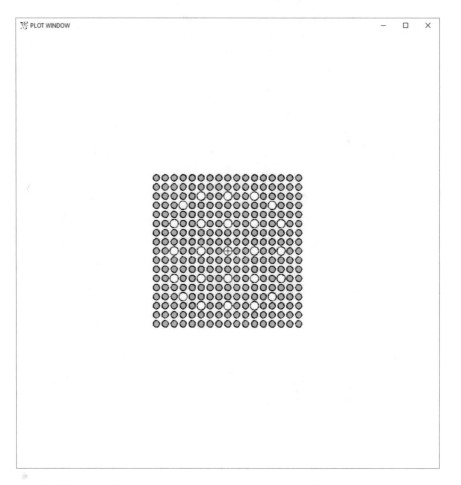

Fig. 4.5 *X–Y* view universe level 2 consisting of universes $u = 1$, fuel (green), and $u = 8$ guide tubes, both lined with zircaloy

and clad are used in this example to illustrate variance reduction for the LSDS problem.

The LSDS source has three overlapping energy distributions to model a thin D-T source made of titanium and deuterium. To provide three different source energy spectra at the point source, the energy is a function of the point position, which is three *x,y,z* triplets repeated three times in distribution SI1/SP1. Then the three different energy distributions—SP3, SP4, and SP5—are given as dependent distributions of the position, FPOS = D2. The source is turned on from $0 < T < 100$ shakes, namely for 1 μs, in distribution SI6.

Tallies are made of the flux in the fuel, F14; flux averaged over all fission chambers, F44; the ^{235}U fission rate, F114; and nubar, F214. For simplicity, fission rate and fission nubar tallies are omitted for ^{232}Th, ^{238}U, ^{239}Pu, and ^{241}Pu. The fission rates and nubar (number of fission neutrons generated) are tallied by multiplying the flux (FM cards) times the fission cross section (reaction MT = −6) and fission nubar (reaction MT = −7). The fission rate and fission nubar tallies have units of fissions and fission neutrons per barn-atom. The standard practice is to use the −1 FM option to multiply by atom density, ρ_a, and multiply by volume, V, by dividing by 1 instead of V with the SD tally card:

```
fm114  -1 5  (-6)
sd114   1
fm214  -1 5  (-6 -7)
sd214   1
```

Tally 114 multiplies the flux/volume tally by $\rho_a \, \varphi \, \sigma_f \, V$ to calculate total number of fissions. Tally 214 multiplies the flux/volume tally by $\rho_a \, \varphi \, \nu \, \sigma_f \, V$ to calculate the total neutrons produced. These quantities are atom density ρ_a, flux φ, nubar ν, fission cross section σ_f, and volume V.

For all tallies, there are 265 logarithmically interpolated time bins from 0 to 20 μs (2e5 shakes):

```
t0   1e2 263ilog 2e5
```

and four energy bins from 0 to 100 MeV:

```
e0   1e-4 1e-3 1e-2 1e2
```

These bins will be printed with five columns of energy bins (including the total energy) and 266 rows of time bins, including the total. The default printing of energies down in columns of time across is overridden by the FQ0 card.

The CUT:N card causes any neutron that survives past 20 μs (200,000 shakes) to be killed so that it is no longer followed in the random walk.

The PRDMP card controls printing and dumping to the RUNTPE file. Setting the third entry to 1 causes an MCTAL tally file to be generated, which is highly recommended for all calculations. The MCTAL file is in ASCII format and can be read across platforms, operating systems, and all MCNP versions. The RUNTPE continuation file contains all the cross sections and necessary information for problem continuation and is generally too large to save. The MCTAL file contains only tallies, is short enough to archive, enables all tallies to be plotted, and can be used for other post-processing functions.

"NPS 1e8" defines the number of particles started and specifies to stop the calculation after 1e8 source histories. These source particles are statistical samples. The physics of the problem is to model a single neutron source that is sampled *NPS* times.

4.1.2.2 Variance Reduction Step 1: Simplify the Problem and Add the Weight Window Generator

The weight window generator is recommended to gain problem insight and possibly provide the needed variance reduction. Plotting the geometry suggests a cylindrical weight window mesh that overlays the geometry. The significant radial changes in the geometry are at radii R = 10.5, 15, 19, 25, 45, 70 cm bounded by a 101 cm cylinder. It is recommended to make the weight window totally enclose the geometry, so a 101 cm outer cylinder will be specified. In the Z-direction, the major boundaries in the geometry are at Z = −50, −30, −15, 15, 30, and 50. Azimuthal angles, widely spaced except at the source, also appear advantageous. Therefore, the following mesh can be defined.

```
mesh geom=rzt origin=0 0 -50 axs 0 0 1 vec 0 1 0 ref -70.6352 0 0
     imesh 10.5 15 19 25 45 70 101
     jmesh 20 35 65 80 100
     kmesh 45 85 95 135 180 240 300 360
```

The above mesh describes a cylindrical mesh overlaying the LSDS (MCNP manual [1] Section 3.3.6.4.4). The mesh is defined as a cylinder (geom = rzt or geom = cyl), with the base of the cylinder at the mesh origin 0,0,−50 and a cylindrical axis in the axs 0 0 1 Z-direction with the azimuth defined normal to the axis in the vec 0 1 0 Y-direction. The reference point is a location in a mesh element that will certainly have particles that eventually score pass through it. Because every particle that scores comes from the source, and because the source is a point, the source point is a good reference point for this problem: ref −70.6352 0 0.

The weight window generator optimizes the calculation of a specified bin in a specific tally. This optimization bin should be one that gets a relative error of 10–20% in just a few minutes so that the weight windows can be iterated. A new tally specifically for optimization needs to be created—in this example, F4. An initial guess for the time resolution is 20 logarithmically spaced times up to 1e5 shakes (20 μs). Only one energy bin is needed to start: E4. The tallies should be printed with the times in a column, (FQ4 card), and the 20th time bin, TF4, is the one of interest.

```
fc4 Average flux in fission chambers - optimize
f4:n   (201  202  203  204  205  206  207  208  209  210
        211  212  213  214  215  216  217  218  219  220
        221  222  223  224  225  226  227  228  229  230
        231  232)
t4 100 18ilog 2e5
e4 100
tf4    7J 20
FQ4 T E
```

The weight window is turned on by specifying the tally to be optimized, F4, and specifying that a MESH will be used.

```
wwg    4  0
```

Once the WWG, MESH, and optimization tally F4 are specified, the weight window generator mesh can be plotted as in Figs. 4.6 and 4.7 and run for a short time by setting

```
NPS    1e5
```

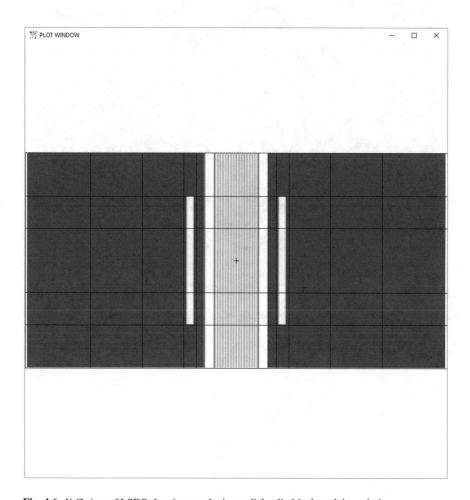

Fig. 4.6 *Y–Z* view of LSDS showing overlaying radial cylindrical mesh boundaries

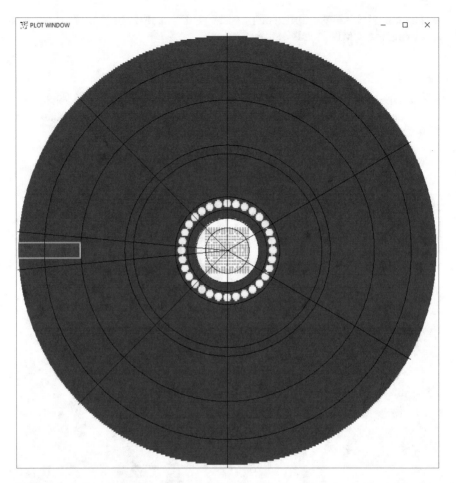

Fig. 4.7 *X–Y* view of LSDS showing overlaying radial cylindrical and azimuthal mesh boundaries

The results of tally 4 in the output file are

```
    time            tally          error
1.0000E+02      7.59815E-05      0.0126
1.4919E+02      5.11286E-05      0.0146
2.2258E+02      3.62428E-05      0.0178
3.3206E+02      2.52853E-05      0.0207
4.9540E+02      1.70135E-05      0.0246
7.3908E+02      1.12966E-05      0.0286
1.1026E+03      7.42396E-06      0.0354
1.6450E+03      5.02155E-06      0.0415
```

```
2.4542E+03    3.48706E-06    0.0498
3.6614E+03    2.15169E-06    0.0630
5.4624E+03    1.39007E-06    0.0810
8.1493E+03    9.38679E-07    0.0973
1.2158E+04    6.88377E-07    0.1322
1.8138E+04    3.23577E-07    0.1790
2.7061E+04    1.24683E-07    0.2150
4.0372E+04    1.08449E-07    0.2554
6.0230E+04    4.76670E-08    0.2620
8.9857E+04    4.61964E-08    0.4296
1.3406E+05    5.08174E-09    0.7642
2.0000E+05    2.46054E-08    0.9450
   total      2.38730E-04    0.0095
```

This information suggests generating weight windows to optimize the 13th time bin of tally 4. The 13th time bin of tally 4 at 12,158 shakes had a relative error of 13.22% and the later time bin errors are too big to generate good weight windows and so the 13th time bin is selected on the tf4 card:

```
tf4    7J 13
```

The weight window generator needs to also generate time-dependent weight windows (MCNP manual [1] Section 3.3.6.4.3):

```
wwgt:n 200 500 1000 1600 3000 8000 1.2e4 2e4 4e4 1e6
```

These bounds were chosen at times when the flux decreased roughly by a factor of 2 according to the previous run, with coarser binning after 12,000 shakes. Sufficient granularity is needed to allow the values to change with the conditions of the simulations. There is more than one successful set of parameters and the values depend on experience and trial and error.

The problem is now run again as:

```
mcnp6   i=inp01   n=j01.
```

To iterate this problem, it is important to link one run to another, which requires careful file naming. The LSDS file "inp01" is the first iteration. The output files will be "j01.*", where "j" is chosen because no MCNP files start with "j." When the run is completed, a one time-bin weight window file, j01.1, and a multigroup time-dependent weight window file, j01.e, are generated. The weight window files contain the description of the weight window mesh to be read and the lower weight window bounds of each mesh.

4.1.2.3   Iteration 2

```
cut:n 200000 J 0 0
wwp:n 5 3 5 0 -1 j j j j
```

cut:n is modified and wwp added and the file now becomes INP02 so that the first iteration, INP01, is retained in case the iteration fails. The CUT:N card adds zero third and fourth entries to use analog capture. It is expected that there will be many late time weight windows that cannot be determined and thus have a value of zero. Histories entering these late time bins which have zero weight windows will then be too quickly killed by implicit capture unless analog capture is invoked. For this reason, the MCNP default implicit capture algorithm is generally incompatible with weight windows. The WWP card tells the MCNP code that

- the upper weight window bounds of each mesh are 5 times the lower bound read from the WWINP file (j01.e). Particles in a mesh with weights above 5 times the lower weight bound will be split in order to have weights in the weight bound range;
- particles that have a weight below the lower bound will undergo Russian roulette, with the survivors having a weight 3 times the lower bound;
- particles will never be split or rouletted more than a factor of 5;
- particles will be checked for weight at all collision and surface crossings (fourth entry = 0); and
- the lower weight window bounds will be read from a WWINP file (fifth entry = −1).

Input file INP02 is now plotted to view the weight window values:

```
mcnp6  i=INP02  n=jp02.  WWINP=j01.e   IP
```

Click

- WWN (right side menu) to make weight windows the edit quantity,
- cell line (lower left menu) to plot the weight window mesh (ww mesh) that is being read,
- L1 sur (lower left menu) to turn off surface labels,
- L2 off (lower left menu) to make edit quantity wwn1:n as the cell label,
- XY (lower left menu) to get the best view.

The plot will show the weight window lower bounds in the first weight window time bin, which is also the source time bin. Note that the lower bound is ~8 in the source mesh cell. Source neutrons will start below the window. At the first collision in the cell they will be rouletted 5 to 1 (third WWP entry). At the next collision in the cell they will be rouletted with a survival weight of 24, which is 3 times the lower weight window bound (second WWP entry). All the windows need to be renormalized by a factor of at least 24 (seventh WWP entry = 0.04) to have source particles start in the mesh weight window.

Clicking the n (lower right menu) will step through the weight window time bins, which can be plotted by double-clicking the L2 button and the Redraw button on the bottom menu bar. For weight windows wwn1:n through wwn7:n, the values of the weight windows drop by about a factor of 2 as desired in both the Z-axis and radial directions. The values are mostly zeroes for the eighth to tenth weight window bins. Because analog capture is turned on, particles will neither split nor roulette when these bins are used at late time in subsequent iterations; the particles will only scatter, be captured, or leak out of the system as if there were no weight windows. The eighth to tenth weight window time bin values are zeroes because tally 4 was optimized for its 13th bin for times up to 12,158 shakes. We also observe that there is not much variation of weight windows in the azimuthal direction, suggesting that fewer azimuthal weight bins are needed.

4.1.2.4   Iteration 2a

```
tf4    7J 20
kmesh 85 95 180 360
wwgt:n 200 500 1000 1600 3000 8000 1.2e4 2e4 4e4 6e4 9e4 1.3e5 1e6
wwp:n 5 3 5 0 -1 j .04 j J
```

The above changes are made to inp02 (which is to be run with the WWINP file j01.e from iteration 2 revised input INP01 with TF4 7J 13windows). It is hoped that the 20th (last) time bin of tally 4 will now have acceptable results when the weight window file from iteration 2 is used, so the target time bin on TF4 is changed. The azimuthal weight window bins are thinned out because they did not appear needed (KMESH). Additional weight window generator time bins are requested (WWGT) corresponding to the late time bins of tally 4. All windows from J01.e are renormalized by a factor of .04 so that the source particles start within the weight window.

```
mcnp6  i=INP02  n=j02.  wwinp=j01.e  IPXR
```

The execution line command IXPR enables plotting INP02 before it is run. The plot (with cells labeled by weight window value) shows that the source particles will indeed enter the weight window (wwn1:n) and, by clicking cell line multiple times, the new mesh with fewer azimuthal divisions is viewed. By clicking end, the plot is finished and now the MCNP code runs 1e5 histories.

Input inp02 is now copied to inp02a. The generated weight windows can be reviewed with the seventh entry on the WWP card set to J:

```
mcnp6  i=inp02a  n=junk.  wwinp=j02.e  IP
```

The generated weight window values in j02.e are not asymmetric in azimuth. The windows increase in some regions with time when they would be expected to decrease because they are more important at later times. Examination of the output file j02.o reveals that the 20th time bin has a 59% relative error—far too large to generate decent weight windows. Also it is much slower because 1e5 source particles generate 1.8e6 split weight window tracks.

Because j02.e is unacceptable, j01.e (iteration 2 optimized) is used again. INP02 is changed by setting optimization to the 16th F4 time bin, which had a smaller relative error and, by keeping the original seventh WWP normalization factor of INP02 (the appropriate value for weight window file j01.e):

```
tf4    7J 16
wwp:n 5 3 5 0 -1 j .04 j j

mcnp6  i=INP02  n=j02.  wwinp=j01.e
```

4.1.2.5 Additional Iterations

Copy inp02 to inp03 and the seventh WWP entry is temporarily set to J to see the unnormalized weight window plots of WWINP = j02.e. These appear to be good, with the generated windows being lower with increasing time and getting closer to the fission chambers. The source enters the window with a lower bound of 45. The 17th energy bin of tally 4 has a relative error of 15%, which indicates that it is good for optimization. (The 18th bin is probably also good.) Renormalize the j02.e windows by a factor of >45 with the seventh WWP entry:

```
tf4    7J 17
wwp:n 5 3 5 0 -1 j .007 j J

mcnp6  i=inp03  n=j03.  wwinp=j02.e
```

Copy inp03 to inp04, set the seventh WWP entry to J, and plot

```
mcnp6 i=inp04 n=jp04. wwinp=j03.e ip
```

The weight window plots of WWINP = j03.e appear to be good:

The source enters the window with a lower bound of about 15 (Fig. 4.8), so setting the seventh entry of WWP to 0.025 would result in a weight window for the source from 0.375 to 1.865, which brackets the source particles of weight 1. The 20th F4 time bin has a value of 3.59e−9 with relative error of 26%, suggesting that it can now be used to generate weight windows. Now change inp04.

```
tf4    7J 20
wwp:n 5 3 5 0 -1 j .025 j J

mcnp6  i=inp04  n=j04.  wwinp=j03.e IPXR
```

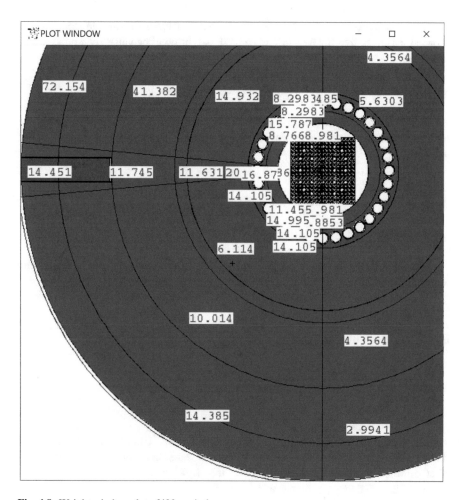

Fig. 4.8 Weight window plot of j03.e windows

The later time bins of tally F4 from output j04.o are

```
    Time            Tally         error
4.0372E+04      1.32776E-07      0.0433
6.0230E+04      8.19473E-08      0.0908
8.9857E+04      5.63077E-08      0.1401
1.3406E+05      1.86379E-08      0.2472
2.0000E+05      4.86418E-09      0.1533
```

These later time bins indicate that the new windows, j04.e, should produce even better weight windows.

Copy inp04 to inp05, set the seventh WWP entry to J (to disable the previous normalization so that the new unnormalized weight window values can be viewed), and plot

```
mcnp6 i=inp05 n=jp05. wwinp=j04.e ip
```

The weight window plots of WWINP = j04.e appear to be good. The source enters the window with a lower bound of about 635, requiring a normalization of 5e−4.

Change

```
wwp:n 5 3 5 0 -1 j 5e-4 j J
```

```
mcnp6 i=inp05 n=j05. wwinp=j04.e ipxr tasks 40
```

Note that tasks 40 causes the MCNP code to run in parallel on 40 threads. The computational time will increase, but the wall clock time will decrease.

The later time bins of tally F4 from output j05.o are

```
     Time           Tally        error
  4.0372E+04     1.25962E-07   0.0334
  6.0230E+04     7.16540E-08   0.0580
  8.9857E+04     5.78666E-08   0.1495
  1.3406E+05     1.92124E-08   0.0874
  2.0000E+05     6.79278E-09   0.0422
```

These later time bins indicate that the new windows, j04.e, should produce even better weight windows. The slope test of the ten statistical checks failed, indicating that another iteration is needed. Also, the 15% error in the 18th time bin, 8.9857E+04 shakes, indicates another iteration.

Copy inp05 to inp06, set the seventh WWP entry to J, and plot

```
mcnp6 i=inp06 n=jp06. wwinp=j05.e ip
```

The weight window plots of WWINP = j05.e appear to be good. The source enters the window with a lower bound of about 482, which requires the same normalization of 5e−4.

Change

```
wwp:n 5 3 5 0 -1 j 5e-4 j J
```

```
mcnp6 i=inp06 n=jp06. wwinp=j05.e ipxr tasks 50
```

All time bins of tally F4 from output j06.o are

```
      Time          Tally      error
  1.0000E+02    7.70505E-05 0.0094
  1.4919E+02    5.19335E-05 0.0102
  2.2258E+02    3.59327E-05 0.0119
  3.3206E+02    2.41735E-05 0.0108
  4.9540E+02    1.65736E-05 0.0113
  7.3908E+02    1.09706E-05 0.0112
  1.1026E+03    7.14615E-06 0.0114
  1.6450E+03    4.62664E-06 0.0117
  2.4542E+03    3.17836E-06 0.0120
  3.6614E+03    2.09733E-06 0.0129
  5.4624E+03    1.41215E-06 0.0135
  8.1493E+03    8.53305E-07 0.0153
  1.2158E+04    5.60799E-07 0.0181
  1.8138E+04    3.42249E-07 0.0180
  2.7061E+04    1.85609E-07 0.0159
  4.0372E+04    1.31680E-07 0.0154
  6.0230E+04    7.31380E-08 0.0209
  8.9857E+04    4.84162E-08 0.0675
  1.3406E+05    1.97451E-08 0.0227
  2.0000E+05    7.10981E-09 0.0175
     total      2.37317E-04 0.0066
```

All ten statistical tests for the 20th time bin of tally F4 now pass. Tally F4 was added to the problem to generate weight windows to transport more late time neutrons to the detector region. Consequently, all the tallies are improved at late time at the expense of a slight degradation of early time results as shown graphically in the next subsection.

4.1.2.6   Cylindrical Mesh Weight Window Summary

The initial input file, INP01, was run for 1e8 histories, which took 11 h and resulted in an FOM for the F4 20th time bin of FOM = 0.98. The variance-reduction run, INP06, with FOM = 45, is 45 times more efficient. The total changes from the initial problem are

```
fc4 Average flux in fission chambers - optimize
f4:n  (201  202  203  204  205  206  207  208  209  210
       211  212  213  214  215  216  217  218  219  220
       221  222  223  224  225  226  227  228  229  230
       231  232)
```

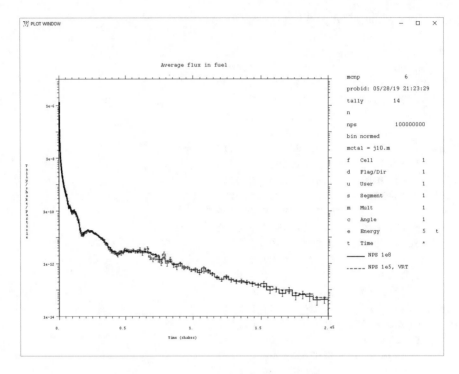

Fig. 4.9 F14 flux in fuel tally results with and without variance reduction

```
t4  100  18ilog 2e5
e4  100
tf4    7J 20
FQ4 T E
c
cut:n 200000 J 0 0
nps    1e5
wwg    4 0
mesh geom=rzt origin=0 0 -50 axs 0 0 1 vec 0 1 0 ref -70.6352 0 0
     imesh 10.5 15 19 25 45 70 101
     jmesh 20 35 65 80 100
     kmesh 85 95 180 360
wwgt:n 200 500 1000 1600 3000 8000 1.2e4 2e4 4e4 6e4 9e4 1.3e5 1e6
wwp:n 5 3 5 0 -1 j 5e-4 j J
```

Figures 4.9, 4.10, 4.11, and 4.12 show the fission chamber tallies both with and without variance reduction. The results are the same, but with variance reduction, the calculation ran 45 times faster to get equivalent results.

The initial 11-h NPS 1e8 run without variance reduction had relative errors ~10% in late time fission (F114) and fission neutron (F214) tallies and relative errors ≪1%

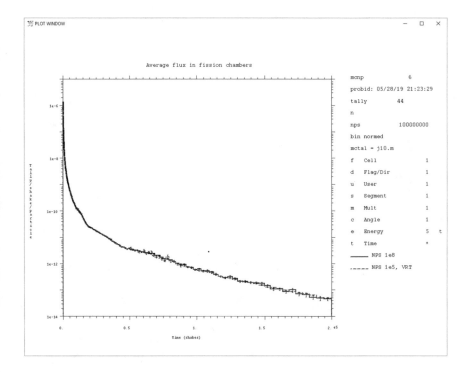

Fig. 4.10 F44 fission chamber flux tally results with and without variance reduction

for early time tallies. The relative errors were 2–5% for all bins of the NPS 1e5 variance-reduction run. Thus, the late time bins benefitted from variance reduction at the expense of the early time bins. The 265 time-bin flux tally likewise had ≪1% error for early time bins and 18% errors for late time bins in the initial run. The range was 1–14% for the variance-reduction run. Thus, the variance-reduction run had good convergence for the 265 time bins tallies and the 20 time bin optimization tally (Figs. 4.9, 4.10, 4.11, and 4.12).

4.1.3 Cell-Based Weight Windows for the Lead Slowing-Down Spectrometer

Example 4.2 provides the changes needed to generate cell-based weight windows for the lead slowing-down spectrometer example of Example 4.1. As with the mesh-based windows, an optimizing tally, F4, that gets good results in a short run is needed to generate the windows. The windows can be generated with or without an initial set of windows (no WWP:n).

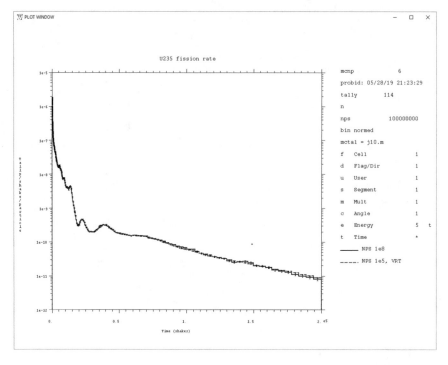

Fig. 4.11 F114 fission chamber fission rate tally results with and without variance reduction

Example 4.2 Lead Slowing-Down Spectrometer Cell-Based Windows Changes to Example 4.1

```
fc4 Average flux in fission chambers - optimize
f4:n   (201 202 203 204 205 206 207 208 209 210
        211 212 213 214 215 216 217 218 219  220
        221 222 223 224 225 226 227 228 229 230
        231 232)
t4 100 18ilog 2e5
e4 100
tf4    7J 20
FQ4 T E
wwg 4 315 .3
wwgt:n 200 500 1000 1600 3000 8000 1.2e4 2e4 4e4 6e4 9e4 1.3e5 1e6
nps 1e5
cut:n 200000 J 0 0
wwp:n 5 3 5 0 0 j 1e-3 j J
wwt:n  2.0000E+02   5.0000E+02   1.0000E+03   1.6000E+03   3.0000E+03
        8.0000E+03   1.2000E+04   2.0000E+04   4.0000E+04   6.0000E+04
        9.0000E+04   1.3000E+05   1.0000E+06
wwe:n  1.0000E+02
```

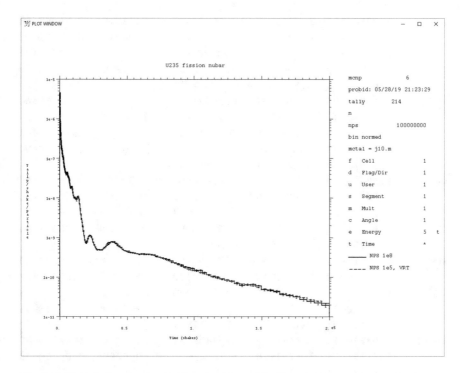

Fig. 4.12 F214 fission chamber fission rate tally results with and without variance reduction

```
wwn1:n   8.8282E+01  1.2704E+02  1.1761E+02  1.1500E+02  1.1481E+02
         1.0430E+02  1.0426E+02  1.0442E+02  0.0000E+00  0.0000E+00
         1.0857E+02  1.0974E+02  0.0000E+00  31R         3.1132E+02
         2.0159E+02  2.3424E+02  3.2016E+02  1.1677E+02  1.2475E+02
         1.6282E+02  2.4031E+02 -1.0000E+00
wwn2:n   2.9654E+01  4.2117E+01  4.2470E+01  3.8329E+01  3.8555E+01
         3.5842E+01  3.6081E+01  3.6670E+01  0.0000E+00  0.0000E+00
         3.1389E+01  3.1733E+01  0.0000E+00  31R         5.9227E+01
         4.4735E+01  4.4908E+01  6.1383E+01  3.8336E+01  3.6095E+01
         3.6931E+01  5.4680E+01 -1.0000E+00
wwn3:n   1.4486E+01  1.9432E+01  2.1266E+01  1.8435E+01  1.8452E+01
         1.7594E+01  1.7747E+01  1.8138E+01  0.0000E+00  0.0000E+00
         1.3747E+01  1.3822E+01  0.0000E+00  31R         3.2168E+01
         2.0333E+01  2.2356E+01  2.9456E+01  1.8042E+01  1.5671E+01
         1.5053E+01  2.0962E+01 -1.0000E+000
wwn4:n   7.6788E+00  1.0647E+01  1.1762E+01  9.5008E+00  9.5196E+00
         1.0351E+01  1.0346E+01  1.0291E+01  0.0000E+00  0.0000E+00
         6.8929E+00  6.9686E+00  0.0000E+00  31R         9.0808E+00
         6.2533E+00  6.0758E+00  9.1151E+00  9.3120E+00  8.0555E+00
         7.3347E+00  9.7978E+00 -1.0000E+00
```

```
wwn5:n   5.9038E+00  8.6340E+00  7.6939E+00  6.9843E+00  6.9996E+00
         7.6940E+00  7.7972E+00  7.6321E+00  0.0000E+00  0.0000E+00
         4.5789E+00  4.6296E+00  0.0000E+00  31R         5.9823E+00
         3.7766E+00  3.5072E+00  5.7816E+00  6.7049E+00  5.2546E+00
         4.5102E+00  5.9946E+00 -1.0000E+00
wwn6:n   4.0332E+00  5.4300E+00  6.0057E+00  4.6442E+00  4.6045E+00
         5.3314E+00  5.3479E+00  5.3253E+00  0.0000E+00  0.0000E+00
         2.5413E+00  2.5524E+00  0.0000E+00  31R         2.7435E+00
         1.8893E+00  2.0213E+00  2.8291E+00  4.3760E+00  2.7616E+00
         2.3227E+00  3.1290E+00 -1.0000E+00
wwn7:n   1.8931E+00  2.9786E+00  2.9564E+00  2.2330E+00  2.1978E+00
         2.5831E+00  2.4478E+00  2.4707E+00  0.0000E+00  0.0000E+00
         1.1206E+00  1.1370E+00  0.0000E+00  31R         1.4655E+00
         1.6101E+00  1.9570E+00  1.6802E+00  2.0827E+00  1.2065E+00
         9.3178E-01  1.2038E+00 -1.0000E+00
wwn8:n   1.4141E+00  1.8389E+00  2.2010E+00  1.6620E+00  1.6507E+00
         1.8549E+00  1.8441E+00  1.8256E+00  0.0000E+00  0.0000E+00
         6.8754E-01  6.8434E-01  0.0000E+00  31R         9.1952E-01
         5.2970E-01  5.6357E-01  1.0146E+00  1.5040E+00  7.3843E-01
         5.7047E-01  7.6280E-01 -1.0000E+00
wwn9:n   4.0374E-01  5.9696E-01  5.9443E-01  4.8738E-01  4.8397E-01
         5.2110E-01  5.1791E-01  5.1250E-01  0.0000E+00  0.0000E+00
         2.8942E-01  2.9186E-01  0.0000E+00  31R         5.6973E-01
         3.4755E-01  4.6019E-01  5.7092E-01  4.7501E-01  3.1770E-01
         2.8366E-01  4.1323E-01 -1.0000E+00
wwn10:n  2.2865E-01  4.9061E-01  3.7397E-01  2.7932E-01  2.7525E-01
         3.5889E-01  3.5108E-01  3.4195E-01  0.0000E+00  0.0000E+00
         1.2345E-01  1.2539E-01  0.0000E+00  31R         2.7895E-01
         1.6249E-01  1.7442E-01  3.2103E-01  2.5545E-01  1.3518E-01
         1.1084E-01  1.7179E-01 -1.0000E+00
wwn11:n  1.6566E-01  2.3132E-01  2.7846E-01  1.4500E-01  1.4157E-01
         4.2491E-01  4.1152E-01  3.7823E-01  0.0000E+00  0.0000E+00
         5.3225E-02  5.3787E-02  0.0000E+00  31R         2.7391E-01
         1.8228E-01  2.1285E-01  3.0000E-01  1.3929E-01  5.4687E-02
         4.9903E-02  9.4521E-02 -1.0000E+00
wwn12:n  2.4018E-02  3.6727E-02  4.7929E-02  2.1757E-02  2.1201E-02
         5.9501E-02  5.7828E-02  5.5140E-02  0.0000E+00  0.0000E+00
         9.5424E-03  9.5949E-03  0.0000E+00  31R         2.5998E-01
         1.6590E-01  2.0152E-01  3.0554E-01  1.8515E-02  1.0339E-02
         1.7634E-02  5.0819E-02 -1.0000E+00
wwn13:n  1.6072E-03  2.3648E-03  2.2318E-03  1.8866E-03  1.9010E-03
         2.1976E-03  2.1924E-03  2.1738E-03  0.0000E+00  0.0000E+00
         1.7331E-03  1.7550E-03  0.0000E+00  31R         1.0552E+00
         3.9534E-01  2.8767E-01  8.1939E-01  2.0065E-03  2.2653E-03
         7.2131E-03  4.7946E-02 -1.0000E+00
```

To generate cell-based windows (starting from nothing or starting from a previous set of cell or mesh-based windows) the WWG card becomes

```
WWG 4 315 .3
```

Cell 315 is the source cell. It is used to normalize the windows. It is chosen because the source cell is certain to have tracks, which eventually score to the target tally. The third entry, .3, sets the lower weight bound in the reference cell, 315, to a lower window bound of .3. From our experience we prefer 0.3 to the default of 0.5 for the third WWG entry to start splitting in important regions sooner.

After one iteration from the mesh-based windows, the newly generated cell-based windows can be used as in Example 4.3.

Example 4.3 Changes to the Lead Slowing-Down Spectrometer Problem of Example 4.1 to Generate and Use Cell-Based Windows

```
fc4 Average flux in fission chambers - optimize
f4:n (201 202 203 204 205 206 207 208 209  210
      211 212 213 214 215 216 217218  219  220
      221 222 223 224  225  226  227  228  229  230
        231  232)
t4 100 18ilog 2e5
e4 100
tf4   7J 20
FQ4 T E
wwg 4 315 .3
wwgt:n 200 500 1000 1600 3000 8000 1.2e4 2e4 4e4 6e4 9e4 1.3e5 1e6
nps 1e5
cut:n 200000 J 0 0
wwp:n 5 3 5 0 -1 j 1e-3 j J
```

The cell-based windows may be read from a WWINP file, as is always the case for mesh-based weight windows. They may also be directly inserted in the input file without the need for WWINP = wwfile on the MCNP execution line, as in Example 4.3. When the windows are in the input file, the "R" notation may be used as in Example 4.3. The "R" notation may not be used if the WWINP file is read by the MCNP execution line: all 31 "0.0000E+00" must be specified. The "R" notation may be used if the WWINP file is read with a READ card in the MCNP input file:

```
READ   FILE=wwfile
```

The F4 tally results are

```
    Time          Tally     error
 1.0000E+02   7.64635E-05 0.0097
 1.4919E+02   5.11358E-05 0.0107
```

```
2.2258E+02    3.60174E-05 0.0124
3.3206E+02    2.39359E-05 0.0108
4.9540E+02    1.63472E-05 0.0113
7.3908E+02    1.11143E-05 0.0115
1.1026E+03    7.26652E-06 0.0121
1.6450E+03    4.62532E-06 0.0121
2.4542E+03    3.18450E-06 0.0122
3.6614E+03    2.11597E-06 0.0132
5.4624E+03    1.38815E-06 0.0133
8.1493E+03    8.42698E-07 0.0153
1.2158E+04    5.47193E-07 0.0156
1.8138E+04    3.44415E-07 0.0157
2.7061E+04    1.86702E-07 0.0152
4.0372E+04    1.28959E-07 0.0154
6.0230E+04    7.13956E-08 0.0168
8.9857E+04    4.99288E-08 0.0207
1.3406E+05    1.97536E-08 0.0179
2.0000E+05    6.99311E-09 0.0167
   total      2.35792E-04 0.0068
```

All statistical tests were passed, and the FOM was 55% higher than when the cylindrical mesh-based weight windows were used in Sect. 4.1.2. This improvement in FOM is because cylindrical mesh-based windows are 55% slower on average than cell-based windows. Rectangular mesh-based windows are also typically 15% slower than cell-based windows.

4.1.4 Time Splitting

Example 4.4 Changes to the Lead Slowing-Down Spectrometer Problem of Example 4.1 to Use Time Splitting

```
fc4 Average flux in fission chambers - optimize
f4:n   (201  202  203  204  205  206  207  208  209  210
        211  212  213  214  215  216  217  218  219  220
        221  222  223  224  225  226  227  228  229  230
        231  232)
t4 100 18ilog 2e5
e4 100
tf4    7J 20
FQ4 T E
nps 1e5
cut:n 200000 J 0 0
TSPLT:n 2 200 2 500 2 1000 2 2000 2 3000 2 6000 2 1e4 2
        2e4 2 3e4 2 6e4 2 1e5 2 1.2e5 2 1.4e5 2 1.6e5 2 1.8e5
```

Example 4.4 provides the changes to the lead slowing-down spectrometer problem of Example 4.1 to illustrate time splitting. As with mesh-based (Sect. 4.1.2) and cell-based (Sect. 4.1.3) weight windows, a target tally is useful to compare results and iterate parameters. Compared with the weight window input files, the generator input (WWG, WWGT) and weight window input (WWP, WWT, WWE, WWN) are replaced by the TSPLT input. The TSPLT card causes a 2-for-1 track split at 200 shakes, at 500 shakes, at 1000 shakes, etc., with the split occurring at the first collision or surface crossing after the time boundary.

The MCNP code does not automatically generate optimum TSPLT parameters like the weight window generator (WWG) does. The user must guess values, entering pairs of split ratio (which may be fractional) and the times at which the splits are to occur. If splitting is insufficient, then there will be no particles that survive to late times. If splitting is excessive, then the problem will run slowly or even hang because too many split particles are followed. The recommended practice is to split sufficiently to keep the relative error roughly constant in the time bins of the target tally. To iterate on the TSPLT values the time bins of the target tally should have the same boundaries as the TSPLT card.

The TSPLT of Example 4.4 passes all statistical tests for the target tally F4:

```
      time
  1.0000E+02    7.72194E-05 0.0127
  1.4919E+02    5.23742E-05 0.0151
  2.2258E+02    3.69117E-05 0.0184
  3.3206E+02    2.48745E-05 0.0182
  4.9540E+02    1.71429E-05 0.0209
  7.3908E+02    1.12507E-05 0.0214
  1.1026E+03    7.19584E-06 0.0233
  1.6450E+03    4.73307E-06 0.0234
  2.4542E+03    3.14349E-06 0.0262
  3.6614E+03    2.08520E-06 0.0260
  5.4624E+03    1.44751E-06 0.0251
  8.1493E+03    8.58901E-07 0.0260
  1.2158E+04    5.64038E-07 0.0273
  1.8138E+04    3.39750E-07 0.0266
  2.7061E+04    1.88853E-07 0.0270
  4.0372E+04    1.26737E-07 0.0264
  6.0230E+04    7.30615E-08 0.0302
  8.9857E+04    5.08255E-08 0.0320
  1.3406E+05    2.13284E-08 0.0412
  2.0000E+05    6.95027E-09 0.0338
     total      2.40609E-04 0.0093
```

Although the TSPLT time splitting for this problem is the easiest method and half as effective as the weight windows, it is effective only because the relative importance of each region of the geometry is the same. If the importance of different regions, different energies, and other aspects of phase space is varied, then simple time splitting will be far less effective.

Table 4.1 Relative performance

Method	FOM 4	FOM 44	FOM 214
No VARIANCE REDUCTION TECHNIQUE	1	1	1
Cyl mesh WW	608	477	1438
Cell WW	1082	592	1964
TSPLT	372	438	1040

The relative performance of the cylindrical mesh-based weight windows (Cyl mesh WW), cell-based weight windows (Cell WW), and time splitting (TSPLT) is shown in Table 4.1.

where

- FOM 4 is the FOM for the $1.3406E+05 < t < 2.0000E+05$ time bin of the F4 target flux tally;
- FOM 44 is the FOM for the $1.9432E+05 < t < 2.0000E+05$ time bin of the F44 target flux tally; and
- FOM 214 is the FOM for the $1.9432E+05 < t < 2.0000E+05$ time bin of the F4 target neutron production tally.

The FOM speedups by these variance-reduction methods range from factors of 372 to 1964 depending on the tally and method, that is, calculations that take 6–33 h can now run in a minute. Alternatively, the resolution is 19–44 times better.

4.1.5 Variance Reduction for the Cf Shuffler

4.1.5.1 Cf Shuffler Modified Input

The Cf shuffler input of Example 3.4 is slightly modified in Example 4.5 to enable exposition of additional variance reduction methods.

Example 4.5 Modified Cf Shuffler Input of Example 3.4

```
vr03.txt for 500 g 235U, density 2.5g/cm^3
C
c       Cells
c
10   5 -2.5   -10        imp:n=1  $ SNM
11   0        -12 10     imp:n=1  $ void about SNM
12   2 -8.65 -11 12      imp:n=1  $ Cd
13   1 -0.94 -13 11 100 101 102 103 104 105 106 107 108
     109 110 111   imp:n=1  $ poly
c
c       3He Tubes
c
```

```
100    3 1.002e-4 -100    imp:n=1 $ #1
101    3 1.002e-4 -101    imp:n=1 $ #2
102    3 1.002e-4 -102    imp:n=1 $ #3
103    3 1.002e-4 -103    imp:n=1 $ #4
104    3 1.002e-4 -104    imp:n=1 $ #5
105    3 1.002e-4 -105    imp:n=1 $ #6
106    3 1.002e-4 -106    imp:n=1 $ #7
107    3 1.002e-4 -107    imp:n=1 $ #8
108    3 1.002e-4 -108    imp:n=1 $ #9
109    3 1.002e-4 -109    imp:n=1 $ #10
110    3 1.002e-4 -110    imp:n=1 $ #11
111    3 1.002e-4 -111    imp:n=1 $ #12
c
9999   0 13        imp:n=0  $ outside cell
c


c
c      Surfaces
c
10 rcc 0 0 7      0 0 7.073553    3      $ SNM
11 rcc 0 0 6      0 0 9           4      $ exterior of Cd
12 rcc 0 0 6.1    0 0 8.8         3.9    $ interior of Cd
13 rcc 0 0 0      0 0 21          14     $ exterior of diagnostic
c
c      3He Tubes
c
100 rcc 8.270    0.000 0.100   0 0 20.8    1.27   $ #1
101 rcc 7.162    4.135 0.100   0 0 20.8    1.27   $ #2
102 rcc 4.135    7.162 0.100   0 0 20.8    1.27   $ #3
103 rcc 0.000    8.270 0.100   0 0 20.8    1.27   $ #4
104 rcc -4.135   7.162 0.100   0 0 20.8    1.27   $ #5
105 rcc -7.162   4.135 0.100   0 0 20.8    1.27   $ #6
106 rcc -8.270   0.000 0.100   0 0 20.8    1.27   $ #7
107 rcc -7.162  -4.135 0.100   0 0 20.8    1.27   $ #8
108 rcc -4.135  -7.162 0.100   0 0 20.8    1.27   $ #9
109 rcc 0.000   -8.270 0.100   0 0 20.8    1.27   $ #10
110 rcc 4.135   -7.162 0.100   0 0 20.8    1.27   $ #11
111 rcc 7.162   -4.135 0.100   0 0 20.8    1.27   $ #12
c


c
c      Materials
m0 nlib=.70c
c
```

```
m1        6000.  1   1001.  2 $ Polyethylene
mt1       poly.10t                    $ S(a,b)
c
m2        48106.  0.0125
          48108.  0.0089
          48110.  0.1249
          48111.  0.128
          48112.  0.2413
          48113.  0.1222
          48114.  0.2873
          48116.  0.0749
c
m3        2003.  1                    $ He-3
c
m5        92235. 1.
c
mode      n
c
c -------------------------------------------------------------------
c    Cf source
phys:n    30 30 0 J J J 0 30 J J J 0 0
sdef pos=d1 tme=d2 erg=d3
sp3       -3 1.175 1.0401  $ Frohner Watt parameters
si1   L   3.8 0 11
sp1       1.0
si2   0e8  10e8    $upper bin boundaries (no option = option H)
sp2   0    1       $ bin probabilities (no option = option D, H, L)
c -------------------------------------------------------------------
c
c      Tallies
fc4    <<<<< Detection Rate per source Cf neutron >>>>>
f4:n   (100 101 102 103 104 105 106 107 108 109 110 111)
fm4       -1 3 103
sd4       1
t4        10e8 98i 1000e8
tf4       7j 7
fq4       t f
c
fc14   <<<<< Variance reduction target tally >>>>>
f14:n  (100 101 102 103 104 105 106 107 108 109 110 111)
fm14      -1 3 103
sd14      1
t14       1e8 8i 10e8 8i 100e8 7i 900e8 950e8 980e8 1000e8 T
tf14      7j 30
fq14      t f
c
```

```
prdmp   j j 1 4 j
print  -30 -162
nps 1e9
ACT   DNBIAS=15
cut:n 1000e8 J 0 0
c
WWG   14 0 .25
WWGT:n 10e8 20e8 40e8 60e8 100e8 8i 1000e8
MESH   GEOM=CYL  ORIGIN=0 0 0  AXS=0 0 1  VEC=0 1 0  REF=3.8 0 11
       IMESH  3 3.9 4 7 9.54 14
       IINTS  3   1 1 1 1    1
       JMESH  6 15 21    JINTS 1  1  1
       KMESH  1              KINTS 1
TSPLT:n 2 1e10   2 2e10   2 3e10   2 4e10   2 5e10   2 6e10   2 7e10   2
8e10   2 9e10
```

The Cf shuffler input of Example 3.4 is repeated in Example 4.5 with the following modifications:

```
m0 nlib=.70c
```

is added to force use of ENDF/B-VII cross sections to enable reproducibility by
utilizing the same data. The TOTNU card is deleted because it has been the MCNP
default for many years. The following target tally is added for variance reduction:

```
c
fc14    <<<<< Variance reduction target tally >>>>>
f14:n   (100 101 102 103 104 105 106 107 108 109 110 111)
fm14    -1 3 103
sd14     1
t14     1e8 8i 10e8 8i 100e8 7i 900e8 950e8 980e8 1000e8 T
tf14    7j 30
fq14    t f
```

The third PRDMP entry is set to generate MCTAL files so that plots of the tallies
can be made across platforms, code versions, and operating systems. Some tables
are turned off with the PRINT card. The requested number of histories is reduced to
a more manageable size. The cut:n card turns off the default implicit capture and
turns on analog capture to prevent implicit capture from working against other
variance-reduction methods.

```
prdmp   j j 1 4 j
print  -30 -162
nps 1e9
cut:n 1000e8 J 0 0
```

4.1.5.2 Particle Production Bias, Time Splitting, and Windows

A few short trial runs using time-splitting guesses and attempts to generate and use weight windows failed; however, these runs made it apparent that nearly all neutrons that produced captures in the ^{3}He detectors after 20e8 shakes (20 s) are from delayed neutrons. Applying delayed neutron bias is done on the activation card:

```
ACT   DNBIAS=15
```

The maximum allowed delayed neutron bias is DNBIAS = 10 (MCNP manual [1] Section 3.3.3.3); therefore, DNBIAS=15 results in up to ten delayed neutron tracks per fission. The MCNP delayed neutron fission bias algorithm reduces the number of prompt fissions to make room for the delayed fissions. Consequently, there are 8.52 delayed neutron tracks per fission and 2.48 prompt fission tracks per fission or 3.44 times as many delayed fission tracks produced as there are prompt fission tracks. The physical values are 2.50 prompt neutrons per fission, 0.016 delayed neutrons per fission, and a delayed neutron fraction of 0.00637.

The delayed neutron bias immediately provided improvement to the FOM by a factor of 100–300 in the 5e10 < time < 8e10 shake (500–800 s) time range, as illustrated in Fig. 4.13.

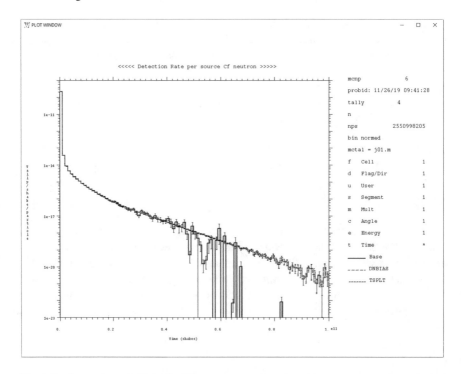

Fig. 4.13 Comparison of Cf shuffler ^{3}He captures as a function of time, comparing the analog run with those using delayed neutron bias and time splitting

An additional factor of 4 improvement in the FOM was achieved with the following time splitting:

```
TSPLT:n 2 1e10   2 2e10   2 3e10   2 4e10   2 5e10   2 6e10   2 7e10   2
8e10   2 9e10
```

Note that the more severe time splits of

```
TSPLT:n 100 10e8   100 20e8   10 30e8   4 60e8   4 1e10   2 3e10
        2 4e10   2 5e10   2 6e10   2 7e10   2 8e10   2 9e10
```

and

```
TSPLT:n 4 2e9 4 3e9 4 6e9 4 10e9 4 15e9 4 22e9 4 32e9 4 43e9
        4 55e9 4 65e9 4 75e10 4 82e9 4 92e9 4 96e9
```

were not as effective.

Cylindrical mesh-based weight windows were also generated.

```
WWG   14 0 .25
WWGT:n 10e8 20e8 40e8 60e8 100e8 8i 1000e8
MESH   GEOM=CYL   ORIGIN=0 0 0   AXS=0 0 1   VEC=0 1 0   REF=3.8 0 11
       IMESH   3 3.9 4 7 9.54 14
       IINTS   3   1 1 1 1     1
       JMESH   6 15 21     JINTS   1   1   1
       KMESH   1                KINTS 1
```

These were used based on tally F14 and two different tally F24s:

```
t24      1e8 8i 10e8 8i 100e8 6i 800e8 1000e8 T
tf24     7j 27
and
t24      1e8 8i 10e8 8i 100e8 4i 600e8 1000e8 T
tf24     7j 25
```

Additional WWGT variations were also tried.

```
WWGT:n 10e8 20e8 40e8 60e8 100e8 6i 800e8 1000e8
WWGT:n 10e8 20e8 40e8 60e8 100e8 4i 600e8 1000e8
```

And different weight window parameter combinations were also tried.

```
WWP:n 5 3 5 0 -1 j 2e-10 j .99
WWP:n 1000 2 5 0 -1 j 1e-12 j .99
```

None of the weight window combinations were found to be effective. In each set of attempts shown, the second was the more favorable.

4.1.5.3 Analysis of Cf Shuffler Variance Reduction

The Cf shuffler problem was made three orders of magnitude more efficient by secondary particle biasing (delayed neutron production) and time splitting. The delayed neutron bias improved the FOM by a factor of 200–300. The time splitting improved the FOM by a factor of 1–4. The net efficiency improvement was a factor of ~1000 at late times (after 6e10 shakes or 10 min) at the expense of efficiency at early times, which had no FOM increase or only a very slight FOM decrease.

Several weight window attempts to get better convergence at late times, $t > 6 \times 10^{10}$ shakes, were unsuccessful. The lack of success is probably because the only way neutrons reached late time was by delayed neutron production and weight windows do not bias physical sampling. Weight windows principally split or Russian roulette particles that enter geometry, energy, or time space.

Furthermore, after the 10×10^8 shake source (10 s), the neutron population falls down ten orders of magnitude by 6×10^{10} shakes (10 min). There are simply too few neutrons present 10 min after the source burst to expect much variance-reduction improvement. Further, the accuracy of the late time tail of the approximate delayed neutron time distribution is poor—many seconds after the fission time.

The factor of 1000 FOM improvement by the application of variance reduction means that a 1 min run can do what would otherwise take 16.7 h or 1000 times fewer parallel processors are needed or time bins can be divided 32 times finer. It is a significant improvement.

4.2 DXTRAN and Other Capabilities for Distributed Source Problems

A UF_6 enrichment cask model provides an example of a distributed source problem. Very little of the source contributes to the desired detector, which often must have very high fidelity because of a large number of small time and/or energy bins. The challenge is to get the important parts of the source to the detector without spending too much computational effort modeling parts of the source that hardly contribute to the needed answer.

Many safeguards problems have very efficient detectors and can be solved with analog methods. Other safeguards problems use Monte Carlo methods that are incompatible with variance reduction, such as coincidence capture; these problems must be analog. Many problems with distributed sources can be analog, but not the following. The UF_6 enrichment cask distributed source problem of Example 4.6 in Sect. 4.2.1 needs variance reduction to get particles to the detector and to undersample most of the source, which contributes very little to the detector.

The distributed source cask problem also showcases a number of MCNP capabilities for understanding problems and optimizing them. Optimization encompasses more than variance-reduction methods to accelerate or improve convergence. There are issues of what physics is needed, what approximations are needed or acceptable, how different MCNP capabilities interact and/or conflict, and good practices. Furthermore, understanding a problem goes beyond modeling the desired tally of an answer and throwing the calculation on a massively parallel computer and hoping for an eventual answer. Problems should be built up in stages, using highly simplified tallies of quantities and other approximations to attain an understanding of the problem. Only then can appropriate methods be applied to get particles where they are needed and to avoid excessive effort in unimportant parts of the problem. The high-fidelity and detailed tallies that are actually desired can then be added after the auxiliary tallies—before using the tallies needed for the final answer. In the UF_6 enrichment cask problem, the F8 pulse-height tally with fine energy bins is the desired tally, but to optimize and understand the problem, F1 current and F6 energy-deposition tallies are used along with next-event estimators.

Massively parallel computers have made it possible to solve many problems by brute force—running them in analog mode without optimization. But optimization is often still important as greater resolution (such as more and finer energy bins), better physics (charged particles, decay, correlated physics models), and larger and more-detailed geometries demand variance reduction. A better understanding of a problem is always desirable.

The capabilities featured here include the following:

- READ cards enable putting sections of the input into separate auxiliary input files so that the entire input file does not have to be repeated just to make small changes.
- FT SCX source distribution tallying to determine the contributions from different source sampling bins
- DXTRAN, a next-event estimator, to

  - direct particles to the detector regions
  - use Detector Diagnostic (DD) Russian roulette to greatly reduce sampling of particles with small probabilities of reaching the detector (MCNP manual [1] Section 3.3.6.11)
  - perform detector diagnostics

- Source position biasing

  - rejection sampling (SDEF CEL or SDEF CCC) enables sampling of irregularly shaped regions
  - fails in some cases with rejection sampling, when, and why

- Mesh-based weight window generation and use

  - proper selection of the MESH REF reference point
  - proper weight window normalization: WWP(7)
  - upper window limits: WWP(9)

- More
  - analog sampling
  - eliminating coherent scatter to improve next-event estimators
  - observations and recommendations

4.2.1   UF₆ Cask Model

The input for the UF_6 cask model consists of a short input file, which uses the MCNP READ capability to include various sections of the input. UF_6 has a phase transition between gas and solid near room temperature. The model has UF_6 solid on the bottom two-thirds of the cask that emits the principal gamma lines of ^{238}U. The top one-third of the cask is gaseous fluorine with no source. The source is sampled uniformly in the volume of the cask and then rejected if not in the UF_6 part on the bottom. The model uses oversimplified HPGe detectors. The goals are to see if these detectors can discern the solid and gaseous regions of the cask and to see the effects of design changes in the HPGe detectors.

Example 4.6 Uranium Enrichment Cask Model. The Master Input File Is "Cask1" with Auxiliary READ Files READ File = Cask_Geom, Cask_Data, Cask_SDEF1, and Tally1
Main input file Cask1

```
UF6 Cask Model
READ File=Cask_Geom
READ File=Cask_Data
NPS   1E6
READ File=Cask_SDEF1
READ File=Cask_Tally1
```

Auxiliary input file Cask_Geom

```
c ***** Cask cells *****
411 111 -5.09  -31        -97              IMP:P=1
412 111 -5.09  -33 31.2 -97                IMP:P=1
413 111 -5.09  -35 31.3 -97                IMP:P=1
416 110 -0.10  -31         97              IMP:P=1
417 110 -0.10  -33 31.2  97                IMP:P=1
418 110 -0.10  -35 31.3  97                IMP:P=1
421 112 -7.874 -32 31                      IMP:P=1
422 112 -7.874 -34 33 32.2 -32.1           IMP:P=1
423 112 -7.874 -36 35 32.3 -32.1           IMP:P=1
431 199 -.001  -40 34 32.2 -32.1           IMP:P=1
```

```
432 199 -.001   -40 36 32.3 -32.1                           IMP:P=1
440 112 -7.874  -41 40                                      IMP:P=1
450 199 -.001   -99 41 #110 #111 #112 #113 #114 #115
                       #116 #117 #118 #119 #120 #121        IMP:P=1
400 0            99                                         IMP:P=0
c ***** Detector cells *****
100   104  -2.700   -202                          U=1       IMP:P=1
101   103  -5.323    202 -203                     U=1       IMP:P=1
102   103  -5.323    203 -204                     U=1       IMP:P=1
103   199  -.001     204 -205                     U=1       IMP:P=1
104   104  -2.700    205 -201                     U=1       IMP:P=1
105   199  -.001     201                          U=1       IMP:P=1
110   199  -.001    -200                      TRCL=300      IMP:P=1
111   Like 110 but TRCL=301
112   Like 110 but TRCL=302
113   Like 110 but TRCL=303 FILL=1
114   Like 110 but TRCL=304 FILL=1
115   Like 110 but TRCL=305 FILL=1
116   Like 110 but TRCL=306 FILL=1
117   Like 110 but TRCL=307 FILL=1
118   Like 110 but TRCL=308 FILL=1
119   Like 110 but TRCL=309 FILL=1
120   Like 110 but TRCL=310
121   Like 110 but TRCL=311

c ***** Cask surfaces *****
31 rcc 0 0   17.73157  0 0 157.8568 36.73
32 rcc 0 0   17.73157  0 0 157.8568 38
33 sph 0 0   142.47 49.53
34 sph 0 0   142.47 50.8
35 sph 0 0    51.07 49.53
36 sph 0 0    51.07 50.8
40 rcc 0 0 0  0 0 207 38
41 rcc 0 0 0  0 0 207 38.4175
97 py 10
99 sph 0 0 0  1000
c ***** Detector surfaces *****
200   SPH   46        0     0      4.7
201   RCC   43.100    0     0      5.427   0     0      3.677
202   RCC   44.527    0     0      4.000   0     0      0.500
203   RCC   43.577    0     0      4.950   0     0      3.200
204   RCC   43.527    0     0      5.000   0     0      3.250
205   RCC   43.227    0     0      5.300   0     0      3.550
```

Auxiliary input file Cask_Data

```
m103 32000 1                            $ Ge
m104 13027 1                            $ Al
M110 19039 1                            $ F
m111 92235 .05  92238 .95  19039 6      $ UF6 5%
m112 26056 1                            $ steel
m199 7014 .8 8016 .2                    $ air
c      x y  z   xx'  yx' zx' xy' yy' zy' xz' yz' zz'
*TR300 0 0 100    0  90 90   90   0 90  90 90 0  $   0 deg from x
*TR301 0 0 100   30 -60 90  120  30 90  90 90 0  $  30 deg from x
*TR302 0 0 100   60 -30 90  150  60 90  90 90 0  $  60 deg from x
*TR303 0 0 100   90   0 90  180  90 90  90 90 0  $  90 deg from x
*TR304 0 0 100  120  30 90  210 120 90  90 90 0  $ 120 deg from x
*TR305 0 0 100  150  60 90  240 150 90  90 90 0  $ 150 deg from x
*TR306 0 0 100  180  90 90  270 180 90  90 90 0  $ 180 deg from x
*TR307 0 0 100  210 120 90  300 210 90  90 90 0  $ 210 deg from x
*TR308 0 0 100  240 150 90  330 240 90  90 90 0  $ 240 deg from x
*TR309 0 0 100  270 180 90    0 270 90  90 90 0  $ 270 deg from x
*TR310 0 0 100  300 210 90   30 300 90  90 90 0  $ 300 deg from x
*TR311 0 0 100  330 240 90   60 330 90  90 90 0  $ 330 deg from x
PRINT -85 -86 -162 -30
PRDMP 2J 1
MODE P
c
```

Auxiliary input file Cask_SDEF1

```
SDEF ERG=D500 POS=0 0 0  AXS=0 0 1  EXT=D501  RAD=D502  CEL=411
SI501 17.73157 175.58837
SP501   0       1
SI502 0 36.73
SP502 -21 1
READ FILE=CASK_SDEF_ERG
```

Auxiliary input file Cask_SDEF_ERG embedded within Cask_SDEF1

```
#       SI500       SP500
         L           D
        1.001030    8.370000e-03
        0.766380    2.940000e-03
        0.742810    8.000000e-04
        0.258230    7.280000e-04
```

```
0.786270    4.850000e-04
1.737730    2.110000e-04
1.831500    1.720000e-04
1.193770    1.347000e-04
1.510500    1.287000e-04
1.434140    9.730000e-05
1.867680    9.180000e-05
1.765440    8.680000e-05
1.875500    8.180000e-05
1.554100    8.080000e-05
1.911170    6.280000e-05
1.237220    5.290000e-05
1.937010    2.890000e-05
1.527210    2.390000e-05
1.414000    2.290000e-05
1.819690    9.000000e-06
```

Auxiliary input file Cask_Tally1

```
FC18 Pulse height in detectors
F18:P ((101 102)<113) ((101 102)<114) ((101 102)<115)
      ((101 102)<116) ((101 102)<117) ((101 102)<118)
      ((101 102)<119) T
E18    .01 198i 2 T
FQ18  E F
```

The combination of these gives the full input file, which is printed in the output file as:

```
 1-    uf6 Cask Model
 2-    read file=Cask_Geom
*********************** begin read ***********************
 3-    c ***** Cask cells *****
 4-    411 111 -5.09   -31       -97          IMP:P=1
 5-    412 111 -5.09   -33 31.2 -97          IMP:P=1
 6-    413 111 -5.09   -35 31.3 -97          IMP:P=1
 7-    416 110 -0.10   -31        97          IMP:P=1
 8-    417 110 -0.10   -33 31.2  97          IMP:P=1
 9-    418 110 -0.10   -35 31.3  97          IMP:P=1
10-    421 112 -7.874  -32 31                 IMP:P=1
11-    422 112 -7.874  -34 33 32.2 -32.1      IMP:P=1
12-    423 112 -7.874  -36 35 32.3 -32.1      IMP:P=1
13-    431 199 -.001   -40 34 32.2 -32.1      IMP:P=1
14-    432 199 -.001   -40 36 32.3 -32.1      IMP:P=1
```

```
15-    440 112 -7.874 -41 40                           IMP:P=1
16-    450 199 -.001  -99 41 #110 #111 #112 #113 #114 #115
17-    #116 #117 #118 #119 #120 #121                    IMP:P=1
18-    400 0              99                            IMP:P=0
19-    c ***** Detector cells *****
20-    100  104  -2.700   -202                U=1      IMP:P=1
21-    101  103  -5.323   202 -203            U=1      IMP:P=1
22-    102  103  -5.323   203 -204            U=1      IMP:P=1
23-    103  199  -.001    204 -205            U=1      IMP:P=1
24-    104  104  -2.700   205 -201            U=1      IMP:P=1
25-    105  199  -.001    201                U=1      IMP:P=1
26-    110  199  -.001    -200        TRCL=300         IMP:P=1
27-    111  Like 110 but TRCL=301
28-    112  Like 110 but TRCL=302
29-    113  Like 110 but TRCL=303 FILL=1
30-    114  Like 110 but TRCL=304 FILL=1
31-    115  Like 110 but TRCL=305 FILL=1
32-    116  Like 110 but TRCL=306 FILL=1
33-    117  Like 110 but TRCL=307 FILL=1
34-    118  Like 110 but TRCL=308 FILL=1
35-    119  Like 110 but TRCL=309 FILL=1
36-    120  Like 110 but TRCL=310
37-    121  Like 110 but TRCL=311
38-
39-    c ***** Cask surfaces *****
40-    31 rcc 0 0    17.73157   0 0 157.8568 36.73
41-    32 rcc 0 0    17.73157   0 0 157.8568 38
42-    33 sph 0 0   142.47 49.53
43-    34 sph 0 0   142.47 50.8
44-    35 sph 0 0    51.07 49.53
45-    36 sph 0 0    51.07 50.8
46-    40 rcc 0 0 0   0 0 207 38
47-    41 rcc 0 0 0   0 0 207 38.4175
48-    97 py 10
49-    99 sph 0 0 0  1000
50-    c ***** Detector surfaces *****
51-    200   SPH  46       0  0  4.7
52-    201   RCC  43.100  0  0  5.427   0  0   3.677
53-    202   RCC  44.527  0  0  4.000   0  0   0.500
54-    203   RCC  43.577  0  0  4.950   0  0   3.200
55-    204   RCC  43.527  0  0  5.000   0  0   3.250
56-    205   RCC  43.227  0  0  5.300   0  0   3.550
*********************** end read ***********************
```

```
  57-
  58-      read file=Cask_Data
************************ begin read ************************
  59-      c m103 32070 .2052    32072 .2745    32073 .0776    32074 .3670
           32076 .0775
  60-      m103 32000 1                          $ Ge
  61-      m104 13027 1                          $ Al
  62-      m110 19039 1                          $ F
  63-      m111 92235 .05   92238 .95   19039 6   $ UF6 5%
  64-      m112 26056 1                          $ steel
  65-      m199 7014 .8 8016 .2                  $ air
  66-      c      x y   z   xx' yx' zx'  xy' yy' zy'  xz' yz' zz'
  67-      *tr300 0 0 100     0  90 90    90   0 90   90 90 0  $   0 deg from x
  68-      *tr301 0 0 100    30 -60 90   120  30 90   90 90 0  $  30 deg from x
  69-      *tr302 0 0 100    60 -30 90   150  60 90   90 90 0  $  60 deg from x
  70-      *tr303 0 0 100    90   0 90   180  90 90   90 90 0  $  90 deg from x
  71-      *tr304 0 0 100   120  30 90   210 120 90   90 90 0  $ 120 deg from x
  72-      *tr305 0 0 100   150  60 90   240 150 90   90 90 0  $ 150 deg from x
  73-      *tr306 0 0 100   180  90 90   270 180 90   90 90 0  $ 180 deg from x
  74-      *tr307 0 0 100   210 120 90   300 210 90   90 90 0  $ 210 deg from x
  75-      *tr308 0 0 100   240 150 90   330 240 90   90 90 0  $ 240 deg from x
  76-      *tr309 0 0 100   270 180 90     0 270 90   90 90 0  $ 270 deg from x
  77-      *tr310 0 0 100   300 210 90    30 300 90   90 90 0  $ 300 deg from x
  78-      *tr311 0 0 100   330 240 90    60 330 90   90 90 0  $ 330 deg from x
  79-      print -85 -86 -162 -30
  80-      prdmp 2J -1
  81-      mode P
  82-      c
************************ end read ************************
  83-      nps  1E6
  84-      read file=Cask_SDEF1
************************ begin read ************************
  85-      sdef ERG=D500 POS=0 0 0  AXS=0 0 1  EXT=D501  RAD=D502  CEL=411
  86-      si501 17.73157 175.58837
  87-      sp501    0        1
  88-      si502 0 36.73
  89-      sp502 -21 1
  90-      read file=CASK_SDEF_ERG
************************ begin read ************************
```

```
 91-      #              SI500        SP500
 92-                     1            D
 93-                     1.001030     8.370000e-03
 94-                     0.766380     2.940000e-03
 95-                     0.742810     8.000000e-04
 96-                     0.258230     7.280000e-04
 97-                     0.786270     4.850000e-04
 98-                     1.737730     2.110000e-04
 99-                     1.831500     1.720000e-04
100-                     1.193770     1.347000e-04
101-                     1.510500     1.287000e-04
102-                     1.434140     9.730000e-05
103-                     1.867680     9.180000e-05
104-                     1.765440     8.680000e-05
105-                     1.875500     8.180000e-05
106-                     1.554100     8.080000e-05
107-                     1.911170     6.280000e-05
108-                     1.237220     5.290000e-05
109-                     1.937010     2.890000e-05
110-                     1.527210     2.390000e-05
111-                     1.414000     2.290000e-05
112-                     1.819690     9.000000e-06
*********************** end read ***********************
*********************** end read ***********************
113-        read file=Cask_Tally1
*********************** begin read ***********************
114-    fc18 Pulse height in detectors
115-    f18:p ((101   102)<113) ((101   102)<114) ((101   102)<115) ((101
            102)<116)
116-    ((101 102)<117) ((101 102)<118) ((101 102)<119) T
117-    e18    .01 198i 2 T
118-    fq18   E F
*********************** end read ***********************
```

The geometry is illustrated in Figs. 4.14 and 4.15.
The plot commands are

```
pz 100 ex 60 la 0 0
```

The plot commands are

```
ba 0 0 1 0 1 0 ex 120 or 0 0 100
```

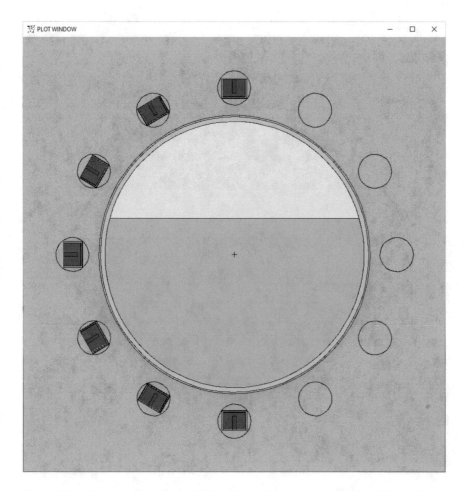

Fig. 4.14 *X–Y* view normal to *Z*-axis of UF$_6$ cask showing 12 detector positions filled with seven detectors. The bottom of the cask is filled with UF$_6$ solid and the top is simply F

There are two primary questions we wish to answer:

- Can HPGe gamma detectors discern that some of the UF$_6$ is solid and the rest is gaseous filling the top part of the cask? A single detector may be rotated around the cask and moved along axially or many detectors may be used. The simulation has 12 detector positions at 30° angles, filled with seven detectors to sample different radial locations.
- What is the effect on the detector of making slight changes to its geometry?

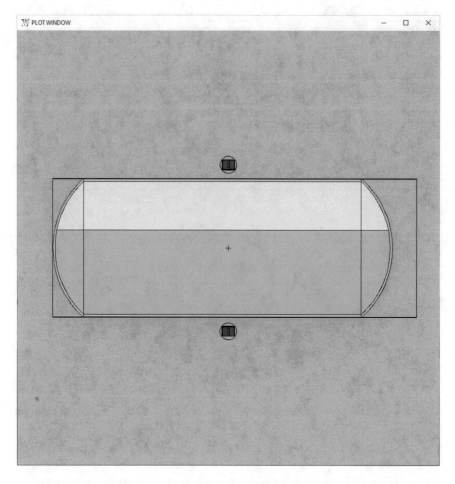

Fig. 4.15 *Z–Y* view of *Z*-axis of UF$_6$ cask, showing top and bottom detectors. The bottom of the cask is filled with UF$_6$ solid and the top is simply F

Most of the contribution to the detectors obviously comes from the source in the UF$_6$ nearest the detectors—that is, the axial region near the detectors at $Z = 100$ cm (center of Fig. 4.15) and the source at the left-side outermost radii (Fig. 4.14). Whereas this source is spread out over 160 cm axially and 37 cm radially, no solution is possible without sampling more of the source near the detectors and directing those source particles toward the detectors—and sampling far less of the source that makes only minimal contributions to the answer. The Cask1 problem was run for hours and no acceptable solution was found to the pulse-height tally. Getting finer energy resolution and seeing small effects in the detector design is impossible.

To understand and optimize the problem requires adding tallies to learn from the MCNP code what is and what is not important. To prepare for a tally that can assess the contributions of various axial regions, the source axial sampling is subdivided by changing SI501 and SP501 to an equivalent distribution:

Auxiliary Input File Cask_SDEF2

```
SDEF ERG=D500 POS=0 0 0  AXS=0 0 1  EXT=D501  RAD=D502   CEL=411
SI501 17.73157   30 5i 90 95 105 110 4i 160 175.58837
SP501   0 12.26843 10 5R  5   10    5 10   4r  15.58837
SI502 0 36.73
SP502 -21 1
READ FILE=CASK_SDEF_ERG
```

Additional tallies are added to the original pulse-height tally. The F101 current tally is most similar to the original pulse-height tally, F18, and tallies the particles that enter and leave a sphere surrounding the 12 detector positions. The F16 tally measures the energy deposition in the seven active HPGe detectors on one side of the cask. The SD changes units to MeV instead of MeV/g. The F116 tally provides the energy deposition in the top, middle, and bottom detectors, subdivided into user bins by FT116 SCX (MCNP Manual [1] Section 3.3.5.18) according to which source axial segment was sampled in distribution SI/SP 501. Note that the F18 pulse-height tally should have had a zero energy bin to identify sampling imbalances (as explained in Sect. 2.4.5).

Auxiliary Input File Cask_Tally2

```
E0      .01 198i 2 T
FC16 Energy deposition in each detector
F16:p ((101 102)<113) ((101 102)<114) ((101 102)<115)
      ((101 102)<116) ((101 102)<117) ((101 102)<118)
      ((101 102)<119) T
SD16  1 7R
TF16 8
FQ16 E F
FC18 Pulse height in bottom detector
F18:P  ((101  102)<113)  ((101  102)<114)  ((101  102)<115)  ((101
102)<116)
      ((101 102)<117) ((101 102)<118) ((101 102)<119) T
E18     .01 198i 2 T
FQ18 E F
TF18 8
FC101 Currents on outer spherical surfaces
F101:P 110200 10i 121200 T
C101 0 1 T
FQ101 C E F
TF101 13 J J J J 1 J J
FC116 Energy deposition in top, middle and bottom detectors
```

```
F116:P ((101 102)<113) ((101 102)<116) ((101 102)<119) T
SD116   1 3R
E116    100
FT116 SCX 501
FQ116 U F
TF116 4
```

When variance reduction is added to improve the convergence of these additional tallies, then these tallies will provide information for understanding and optimizing the problem.

General Recommendations fot the Application of Variance Reduction Techniques:

- Modify the problem into simpler problems that can run fast and provide an understanding of the problem. Such simplification may entail the following:

  - Eliminating or adding tallies

    Eliminating the tallies of interest and using ones to answer specific questions. In particular, use point detectors or DXTRAN to get answers in specific locations.

    Adding tallies to provide specific information. The FT SCX and SCD special tally treatments enable the user to learn which source distributions contribute to the tallies of interest. The tally tagging, FT TAG, cell and surface flagging, CF and SF, tally segmenting, FS, and others contribute to information about which parts of the problem are important.

    Reducing the number of tally time, energy, or other bins so that adequate answers are achieved quicker in the broader bins. A 4-keV energy bin will have twice the efficiency (FOM) of a 1-keV bin.

    The simpler problems used for understanding should omit the F8 tally, which slows down the calculation, particularly when DXTRAN is used (typically 50% slowdown in our experience). The F8 tally is left in these examples for comparison and completeness, not as a best practice for optimization and understanding.

  - Simplifying the source to one characteristic of the problem but less detailed than the one ultimately desired.
  - Adjusting problem physics for optimization studies:

    Thick-target bremsstrahlung affects tallies with energy $E < 100$ keV in this UF_6 cask problem. If that region is not of interest, the second PHYS:P entry should be used to turn it off. If bremsstrahlung is important, then full electron transport should be used, and perhaps bremsstrahlung biasing (PHYS:E) should also be used.

    If electrons are used, turning knock-ons off (PHYS:E) usually does not affect results and speeds electron transport by about a factor of 7.

    In charged particle calculations, scoping studies can be performed faster by eliminating some of the particle types (MODE), particularly ion recoils (PHYS:N).

Simpler materials with fewer isotopes—particularly trace amounts—can be used. Note the simplification of uranium, Ge, iron, air, etc., in this UF_6 cask model.

- Use READ auxiliary input files. Often MCNP input files are thousands of lines long and can be made more manageable by putting parts of the problem that do not change into separate files. Put parts of the input that change into small files. In this UF_6 example, the geometry, materials, transformations, and source energy remain unchanged and are in separate auxiliary input files. The number of histories can be easily modified in the short master file.
- Optimization of the simplified problem can then be applied to the original problem. The insights gained from the simple problem direct the final variance reduction of the original problem. In the case of a superimposed weight window mesh, the simplified problem mesh can be used directly in the original problem.

4.2.2   DXTRAN

The DXTRAN technique is typically useful when a small tally region is inadequately sampled due to a very small probability of particles being transported toward that area. It is among the most effective collision directional biasing variance reduction technique for neutrons and photons available in the MCNP code.

In this example, DXTRAN introduces a so-called DXTRAN sphere that encloses the tally region. At the source and at every subsequent collision reaction that the particle undergoes in its random walk, a special DXTRAN particle is created. This DXTRAN particle is deterministically transported to the surface of the DXTRAN sphere. The random walk of the parent particle is unchanged (other than the random number sequence needed to generate the DXTRAN particle). In particular, the parent particle weight is not reduced by the created DXTRAN particle. The weight of the created DXTRAN particle is offset by killing the parent particle if it reaches the DXTRAN sphere during its random walk. Thus, the DXTRAN event may be thought of in two ways: It may be considered as a splitting of particles into parts that enter the DXTRAN sphere on the next event and particles that do not enter the DXTRAN sphere on the next event. Alternatively, the DXTRAN event may be considered as an accounting process in which weight is both created and destroyed on the surface of the DXTRAN sphere.

DXTRAN puts neutrons or photons on a sphere surrounding a region of interest by using a next-event estimator similar to the point detector. The F5 point detector accumulates the flux, Φ, at a point by a contribution of

$$\Phi = Wp(\mu)\frac{e^{-\sigma x}}{2\pi R^2}.$$

This estimate of Φ is accumulated at every source or collision event.

W = particle weight
$p(\mu) = p(\Omega)/2$ = scattering probability density function
R = distance between event and detector point
$e^{-\sigma x}$ = attenuation through all cells/materials along path R

Although the probability of scattering exactly in the direction, Ω, from the event to the detector is zero, the probability density function of the cosine, μ, assuming azimuthal symmetry, $p(\mu) = p(\Omega)/2$, is finite:

$$0 \le p(\mu) < \infty$$

$$-1 \le \mu \le 1$$

For photon coherent scattering with form factors, $p(\mu)$ is nearly a delta function at $\mu = 1$:

$$p(1) \approx \infty.$$

Consequently, convergence is poor unless coherent scatter is turned off (PHYS:P 2J 1) or the special treatment, FT5 PDS, is applied to sample collision outcomes rather than incident reaction probabilities. (Coherent scattering can cause large weight contributions to next-event estimators and since DXTRAN spheres are a bigger target than point detectors the problem is exacerbated with DXTRAN.)

For DXTRAN, particles are put on the DXTRAN sphere with the weight of the expected value of their being scattered to the sphere:

$$N = W \cdot p(\mu) \cdot V \cdot e^{-\sigma x}$$

$$V = 1 - (R^2 - D^2)^{\frac{1}{2}} / R$$

where

D = radius of (outer) DXTRAN sphere;
R = distance from source/collision to the center of the DXTRAN sphere.

The next-event method samples the contributions to the DXTRAN sphere(s) at every source or collision event using an essential built-in variance-reduction method of Russian roulette. At each source or collision event, only the attenuation to the DXTRAN sphere, $e^{-\sigma x}$, is unknown, and $e^{-\sigma x} \le 1$ always. The remaining terms of the DXTRAN contribution are

$$N > WVp(\mu) < 0.1N_a.$$

Here, N_a is the average contribution from all the previous histories run so far. If

$$N < 0.1N_a$$

Russian roulette is played on these contributions with survival probability $N/(0.1 \times N_a)$ and survival weight $0.1 \times N_a$, which usually results in about one transmission to the DXTRAN sphere for each source history rather than many insignificant contributions from each source or collision event. But because the Russian roulette criteria of N_a is different for each history, the central limit theorem of statistics is violated because all samples are no longer drawn from the same distribution. The effect of this outcome is insignificant in our experience. But another consequence is that histories cannot be rerun with the RAND HIST (or DBCN eighth entry) capability to determine with an event log (DBCN third and fourth entries) to understand the cause of large DXTRAN contributions, which need to be sampled more to achieve good convergence. Therefore, it is recommended that the Russian roulette criteria for each detector be set with the DD card.

The variance-reduction parameters for DXTRAN are added in yet another auxiliary READ file, Cask_VRT2:

```
PHYS:P 2J 1
CUT:P 2J 0 0
DXT:P    0  46 100 4.8 4.8
       -46   0 100 4.8 4.8
         0 -46 100 4.8 4.8
         2e-6 1e-6
```

The PHYS:P physics data use the third entry to turn off coherent scatter. If coherent scatter is left on, then the density function for scattering term, $p(\mu)$, becomes very large when scattering sends the particle nearly forward. Without turning scattering off, the tally convergence is nearly always poor with slope <3 and FOM much lower. Turning off coherent scatter is generally an excellent approximation. To check if turning off coherent scattering is a good approximation, a short run (after variance reduction is completed) should be made with and without coherent scatter to see if the tallies agree within statistics. If coherent scattering is important, then nested DXTRAN, described shortly, may be required.

The CUT:P cutoffs force analog capture. Photon transport is generally analog. There is no weight reduction by capture. Photoelectric absorption results in fluorescence, which gives zero, one, or two photons; however, when variance reduction, such as DXTRAN, position biasing, bremsstrahlung biasing, etc., is used, there are many weight fluctuations. The weight cutoff of implicit capture will preferentially roulette low-weight important particles and not roulette high-weight unimportant particles. Analog capture generally should be used with variance reduction.

The DXTRAN spheres are specified with DXT:P in quintuples for each DXTRAN sphere, followed by two weight cutoffs. The three spheres are at x, y, z = 0 46 100 top; x, y, z = −46 0 100 middle; and x, y, z = 0 −46 100 bottom. Each sphere has an inner and outer radius. The first and second radii are the inner and outer. The second (outer) radius is where the next-event DXTRAN particles are placed. The first (inner) radius should surround the region of interest and is used to aim 80% of the particles placed on the outer sphere towards the inner sphere. When the surrounding medium is not much of a scattering medium—as in this problem, where it is air— the two radii are generally the same. When the DXTRAN sphere is in a scattering medium, then nested DXTRAN spheres should be used where the inner sphere is at radius D; then the outer radius would be the lesser of D + mfp or $2D$, where mfp is the mean free path. For nested spheres, each inner and outer radius would both be D and nesting would be twice the radius of the outer sphere, namely D, $2D$, $4D$, $8D$, ..., out to a radius of D + mfp, because the weight of particles placed on the sphere falls off as $1/D^2$ even in a vacuum and $e^{-\sigma x}$, where σ = 1/mfp.

The DXT weight cutoffs, 2e−6 1e−6, are important when there is more than one DXTRAN sphere so that the DXTRAN spheres do not "talk to each other"; that is, particles in one sphere collide and create particles on the other spheres, which then create more on the others. The second cutoff, 1e−6, is the weight below which Russian roulette is invoked. The first, 2e−6, is the survival weight. Thus a particle in the DXTRAN sphere with weight 5e−7 will survive with weight 2e−6 a quarter of the time and be killed the rest of the time. The DXT weight cutoff should be 10% or lower than the lowest average weight reaching any of the DXTRAN spheres (determined by short test runs). If the cutoffs are absent, there will be the warning message:

```
Warning. More than ONE photon sphere and no weight cutoff.
```

The particles will be followed until their weight drops below the smallest number allowed in the computer. The DXTRAN diagnostics, Print Table 150, will show those transmissions as being killed by "underflow in transmission." These transmissions will be costly in terms of FOM efficiency. When the weight cutoffs are specified, the transmissions killed will show up under "weight cutoff" in the DXTRAN diagnostics, print Table 150, but not in "weight cutoff" in the problem summary table.

It is a good practice to fix the Russian roulette criteria for next-event estimators using the DD card. DD2 refers to photon DXTRAN spheres. DD1 refers to neutron DXTRAN spheres. DD115 would refer to point detector F115. DD0, or just DD, applies to all. For this problem, a good first guess was:

```
DD2    -2e-6 1e5    -2e-6 1e5    -2e-6 1e5
```

The negative sign makes the value absolute rather then relative to the fluctuating average value. The DD Russian roulette criteria for each sphere are $N = 2e-6$. Any transmission to the DXTRAN sphere that puts a particle on the outer sphere with a weight greater than $1e5 \times 2e-6$ will have a diagnostic printout between the first 50 histories (print table 110) and the problem summary table in the output OUTP file. These diagnostics provide all the parameters of the DXTRAN transmission:

```
idx  dx wgt   col wgt  psc    amfp   dd    erg   cell   nps   ncp  p   2
2.5442E-01 1.0000E+00 9.2550E+00 3.0041E-01 1.3708E+01 1.9927E-01
440 2092167 0
```

IDX	= 2 is the DXTRAN sphere number;
DX WGT	= 2.5442E−01 = weight of particle put on DXTRAN outer sphere;
COL WGT	= 1.0000E+00 = weight of particle at source or collision;
PSC	= value of $p(\mu)$, which is 0.5 for pair production and photoelectric and can be >1.0 for incoherent scatter even after turning off coherent scatter as in this case where it is 9.2550E+00;
AMFP	= 3.0041E−01 = average mean free path;
DD	= 1.3708E+01 = distance from source/collision to outer DXTRAN sphere;
ERG	= 1.9927E−01 = energy of particle at source or collision;
CELL	= 440 = cell from where particle is starting; in this case 440, the steel case around the cask;
NPS	= 2092167 = history number, which can be used with RAND HIST or DBCN eighth entry to re-sample the history if DD Russian roulette control is set;
NCP	= 0 = number of previous collisions;
P	= particle is a photon.

A good choice for DD Russian roulette criteria is a tenth of the DXTRAN particle "average weight per history" in the DXTRAN diagnostics Print Table 150, which can be obtained from a short run. For this problem, the choice would be

```
DD2   -8.1e-6 1e4   -9.8e-6 1e4   -10.0e-6 1e4
```

Short runs for this problem indicated insensitivity to the choice of DD parameters. The FOM was about the same for the above Roulette criteria, the 2e−6 criteria, and omitting the DD card altogether. In other problems, the choice can be significant.

The problem can be run by changing the main input file:

Main Input File Cask2

```
UF6 Cask Model
READ File=Cask_Geom
READ File=Cask_Data
NPS  1E7
READ File=Cask_SDEF2
READ File=Cask_Tally2
READ File=Cask_VRT2
```

When DXTRAN is used in this problem, the MCNP code issues the two warnings:

```
warning.  f8 variance reduction has not been verified for more than
one dxtran sphere
```

Whereas the answers appear to agree with the runs without DXTRAN, we believe that, in this case, the F8 pulse-height tallies with DXTRAN are correct, but they must be used with caution and should be verified against a later run with a single DXTRAN sphere or no DXTRAN if a no-DXTRAN run can have good enough convergence for comparison.

The second warning is

```
warning.  tally partly inside and outside of a dxtran sphere.
```

Tallies on surfaces or in cells that are both in and out of a DXTRAN sphere have either false convergence or poor convergence (both indicated by slope <3) because scores outside the DXTRAN sphere are rare, with large weight; scores inside the DXTRAN sphere are numerous, with low weight. In other words, the part of the tally outside the DXTRAN sphere is undersampled relative to the part inside. For this cask problem, if the warning is understood, then it is determined not to be a problem. In this case the warning is understood. The scores in the individual cells (F16, F18, F116 tallies) or spherical surfaces (F101 tally) are correct. The total bin for these tallies includes cells or surfaces that are both in and out of the three DXTRAN spheres and consequently the total bin tally may converge poorly and not pass the convergence tests.

All MCNP warning errors must be reviewed and understood to have full confidence in a calculation.

In addition to the DD detector diagnostics and Russian roulette capability, there is another DXTRAN option that is not used here: the DXC DXTRAN cell contribution capability. Cells may be selected to contribute to the DXTRAN spheres with probabilities $0 < P < 1$ rather than attempting to send every source or collision particle to the DXTRAN spheres. The DXC capability is exactly like the PD capability for point detectors.

The DXTRAN problem with main file Cask2 and VARIANCE REDUCTION TECHNIQUE auxiliary input file Cask_VRT2 provided great problem insight in short runs. This run—and slight variations of—it led to the following:

Observations
- The tally from the three different HPGe detectors in the three DXTRAN spheres was sufficiently different to determine the differences in the shape of UF_6 and F in the cask.
- Using DXTRAN improved the FOM by more than a factor of 7.
- Turning off coherent scatter (PHYS:P 2J 1) increased efficiency by 50% and improved the convergence slope.
- Using three DXTRAN spheres was more than a factor of 2 more expensive than using only one.
- Eliminating the F8 pulse-height tally improved the FOM by 10%.
- Efficiency was not very sensitive to the DD Russian roulette criteria.
- Including the weight cutoffs on the DXT is essential when there are multiple DXTRAN spheres.
- F8 pulse-height tallies must be used with caution if there are multiple DXTRAN spheres.

Recommendations
- Use F5 point detectors or DXTRAN to gain problem insight.
- Use DXTRAN to improve efficiency if the objective is to sample a large source region going to a small tally region:

  - Turn off coherent scatter for photon DXTRAN.
  - Use analog capture, CUT, if weight windows are not used.
  - Have weight cutoffs on the DXT if there are multiple DXTRAN spheres to reduce cross talk between detectors.
  - Use nested DXTRAN if the DXTRAN spheres are in a scattering medium and there are collisions near the DXTRAN sphere, choosing the sphere radii factors of two larger with each nesting.
  - Limit the number of DXTRAN spheres. It may be better to run a separate problem rather than have several DXTRAN spheres at multiple locations.

- The DXC cell contribution should be considered when cells have many collisions but few get to the DXTRAN sphere. It is particularly important when there is a large shield or zero-importance region between the cell and the DXTRAN spheres. In these regions DXC should be set to zero in order not to waste computational time (Ref. [3] Section 12.8). DXTRAN should not be used for deep penetration.
- Use appropriate DD parameters so that the DXTRAN Russian roulette uses the same criteria for all histories and so that detector diagnostics for high scores are neither too many nor too few.
- Be sure to understand all warning messages.

4.2.3   Source Position Biasing

The source position of this UF_6 cask problem is already biased by rejection. The cylindrical volume is uniformly sampled and then positions in the F gas above the UF_6 solid are rejected as source positions. But further biasing is needed. As both speculated and demonstrated by short runs, the positions that contribute most to the HPGe detectors are those close to it. Positions deep within the UF_6 that have a small radius likely do not contribute much. Positions axially far from the detectors at $Z = 100$ also have smaller contributions, which waste time. Thus, both radial and axial biasing are desired.

The following radial sampling was applied, with the probabilities being the proportional volumes:

```
SDEF ERG=D500 POS=0 0 0   AXS=0 0 1   EXT=D501   RAD=D502   CEL=411
SI502 0 10 20 25 30 32 34 35 36 36.73
SP502 0 100 300 225 275 124 132 69 71 53.093
```

with the following tally added to determine the relative contributions of the radial bins:

```
FC126 Energy deposition in top, middle and bottom detectors
F126:P ((101 102)<113) ((101 102)<116) ((101 102)<119) T
SD126 1 3R
E126   100
FT126 SCX 502
FQ126 U F
TF126 4
```

This tally showed the relative contributions of the different radial bins of SI502/ SP502. Most of the tally to the top HPGe detector came from the $R < 20$ inner radial bins. Most of the contributions to the bottom HPGe detector came from the outer radial bins, with less than 0.5% coming from $R < 25$ cm. Thus, if radial biasing is used, it would have to be a separate bias for each detector.

Source radial biasing was then applied using the relative contributions from tally F126 to the bottom detector:

```
SI502 0 10 20 25 30 32 34 35 36 36.73
SP502 0 100 300 225 275 124 132 69 71 53.093
SB502 0 1   1   2   8   8    10   10 15 20
```

The MCNP code issued the following warning:

```
Warning. Biased source position rejection sampling is dangerous.
```

The warning necessitates running the problem with and without biased source position sampling. Indeed, the radial biasing gives wrong answers. The results with and without the SB502 bias were significantly different for all tallies. If the tallies are not sufficiently converged in a short run, then a testing tally should be devised that converges quickly. The reason why source radial biasing with rejection sampling does not work in this problem is that sources in the innermost radial position $R < 10$ are never rejected and sources preferentially biased to outer radial positions are more often rejected; that is, once the radial source position is sampled, the rejection scheme is different for each sampled position, rejecting 33% of the source points for $R > 37$ cm and none of the source points for $R < 10$ cm.

Axial biasing can be applied to the SI501/SP501 distribution using information from the F116 tally with FT116 SCX 501. For the top detector, 3% of the tally comes from the two end regions, which suggests that not sampling the rounded end caps of the cask is insufficient for the top detector and different bias is needed for the top detector. For the middle detector and the bottom detector, the contributions from the different axial regions are nearly the same. Only 1.5% of the contribution comes from the end segments on either side; that is, from $17.73 < Z < 50$ cm and from $150 < Z < 175.6$, which suggests a bias of

```
SI501 17.73157   30 5i 90 95 105 110 4i 160 175.58837
SP501   0 12.26843 10 5R  5   10   5 10   4r  15.58837
SB501 0 1 1 1 1 2 7 24 35 100 35 24 7 2 1 1 1
```

Applying this axial bias again resulted in the warning that "Biased source position rejection sampling is dangerous." But now the answers agree well with the unbiased answers. The axial bias works with rejection because, once the position is chosen, the rejection is the same in each axial section. (Radial bias is discarded for the rest of this problem.)

A good bias for the top position can be inferred from the FT116 SCX 501 tally which gives the tally contribution from each source region. The axial source bias should then be proportional to the tally values for the top detector position:

```
SB501 0 6 7 14 21 36 56 76 47 100 47 76 56 36 21 14 7
```

Why are the values on the SB501 distribution not increasing toward the center? The tally contribution—and hence the recommended bias—increases from 7 to 100. Although the values on the SB501 are not monotonically increasing toward the center, the source bias is: the dip from 76 to 47 to 100 is because the SP501 axial bin source probability size changes.

The SB501 only works for the lower detectors. The fact that the end segments contribute 7% as much as the middle segment indicates that the spherical end caps of the cask that have the source omitted should be considered. But including them results in rejection on a different shape than in the cylindrical part of the cask and would therefore be incorrect. Consequently, an axial bias for the top detectors is approximate and not strong enough to be much advantage; the source biasing for the top detectors should not have axial bias applied.

The above axial bias improves the convergence efficiency for the F116 energy-deposition tally in the bottom position from FOM 2180 to 3170 or by 45%; however, this axial bias decreases the F116 energy-deposition tally in the top position from 1760 to 340, a factor of 5.

The detectors are, relative to the X-axis, at 90° (top), 120°, 150°, 180°, 210°, 240°, and 270° (bottom). The DXTRAN spheres are at 90°, 180°, and 270°. If source position biasing is used, then a radial biasing fails, and axial biasing must be different for the top and bottom regions and perhaps different for the region around 150° opposite the UF_6 and F gas interface.

Observations
- Radial position biasing fails with rejection position sampling. The only indication is a warning message and then subsequent comparison to a run without the radial position biasing rejection sampling combination.
- Axial position biasing works with rejection position biasing, but for this problem, the optimum axial bias is significantly different for the top and bottom detector tallies.
- The spherical end caps should be included when calculating the contribution to the top detectors.
- Axial biasing is of benefit for calculation of the bottom detectors.

Recommendations
- Analog capture should be used so that low-weight source particles are not rouletted by the default MCNP weight cutoff. The low-weight particles are the ones biased to be the most numerous and important.
- Do not use source radial biasing for this problem because it interacts incorrectly with source position rejection sampling.
- Axial biasing requires dividing the problem into upper, bottom, and perhaps middle detector problems.

  - Do not use axial biasing for the top detectors of this problem.
  - Axial biasing is advantageous for the bottom detectors of this problem.

- Be sure to understand all warning messages.

4.2.4   Best Single Detector Solution

The UF_6 cask problem can best be solved for each detector individually with a single DXTRAN sphere, axial source biasing, and weight windows tailored to the specific detector. The case for the bottom detector is presented, but the same approach would be used for any other detectors.

The problem can be further optimized by either mesh-based weight windows or importances. Cell-based weight windows are counterproductive because the importance of different regions of the large cells varies extremely. Mesh-based weight windows can have meshes small enough so the lower weight bounds vary by only a factor of 4 or so between mesh cells.

Although they are not used in the following example, cell importances would also be valuable. The importances are ignored within the weight window mesh for nonzero mesh values. In this UF_6 cask problem, without the weight window mesh, the importance should increase as particles transition from the UF_6 (cell 411) or F gas (cell 412) into the first (cell 421) and second (cell 440) layers of the iron case, which can be achieved by assigning importances as:

```
Cell              IMP:P
411               0.125
412               0.125
421               0.25
440               0.5
Others            1.0
```

or

```
Cell              IMP:P
411               1
412               1
421               2
440               4
Others            8
```

If these importances were implemented, they would be entered in the auxiliary Cask_Geom file. This would be case Cask3 which is neither shown nor further described.

A mesh-based weight window with solution is now presented as case Cask4. The Master file is:

```
UF6 Cask Model
READ File=Cask_Geom
READ File=Cask_Data
NPS  1E7
READ File=Cask_SDEF4
READ File=Cask_Tally4
READ File=Cask_VRT4
```

The auxiliary input files Cask_Geom and Cask_Data and Cask_SDEF_ERG are unchanged. The new files are

Auxiliary Input File Cask_SDEF4

```
SDEF ERG=D500 POS=0 0 0   AXS=0 0 1   EXT=D501   RAD=D502   CEL=411
SI501 17.73157   30 5i 90 95 105 110 4i 160 175.58837
SP501   0 12.26843 10 5R  5   10    5 10   4r   15.58837
SB501   0 1 1 5 16 60 28 1000 2000 1020 280 60 9 3 1 1 1
SI502 0 36.73
SP502 -21 1
READ FILE=CASK_SDEF_ERG
```

The values on the SB501 axial source bias were chosen to be inversely proportional to the generated weight windows for the portion of the cask closest to the detector. See the values of the windows in Fig. 4.16. (Because the volumes of the regions change, the jump in particles per unit volume is not as great as it appears.) This selection enables particles that exit the case to enter weight windows that match the source axial bias. The source weight range is printed below the summary table in the output file:

```
range of sampled source weights = 7.1045E-02 to 4.4299E+02
```

Auxiliary Input File Cask_Tally4

```
FC16 Energy deposition in bottom detector
F16:p   ((101 102)<119)
E0      .01 198i 2 T
SD16   1
FC18 Pulse height in bottom detector
F18:P ((101 102)<119)
E18     0 1e-5 .01 198i 2 7I 10 T
FC101 Currents on outer spherical surfaces
F101:P 119200
C101 0 1 T
FQ101 E C
FC116 Energy deposition in bottom detector
F116:P ((101 102)<119)
SD116 1
E116   100
FT116 SCX 501
FQ116 U F
```

The weight windows are generated by iterating on the weight window mesh in Cask_VRT4, which also uses DXTRAN with appropriate weight cutoffs (DXT) and Russian roulette criteria (DD2). The seventh WWP parameter, 0.004, is chosen so that the minimum source weight from the source axial bias, 0.071, at the cask lining

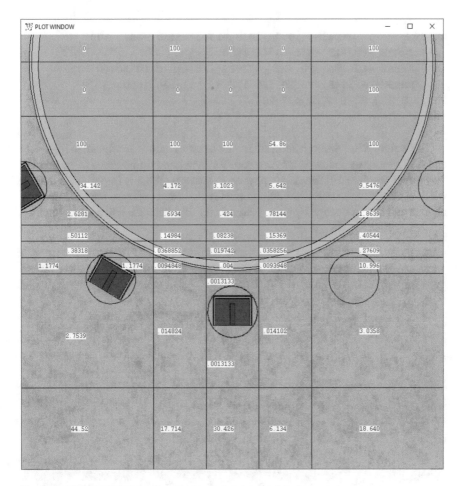

Fig. 4.16 Weight window mesh normal to cask axis near detector

closest to the detector enters a weight window mesh with lower bound 0.004. The upper bound is 5 times higher (first entry on WWP), so the source particles closest to the detector split five for one. Most source particles far from the detector enter a weight window with lower bound 100, which is the ninth entry on the WWP card; or they enter windows with lower bound 0, which causes them to avoid weight window splitting/roulette.

Auxiliary Input File Cask_VRT4 (Note: For the First Run, the WWP:P Card Should Be Commented Out)

```
PHYS:P 2J 1
CUT:P 2J 0 0
DXT:P   0 -46 100 4.8 4.8
        2e-4 1e-4
```

```
WWG 16 0 1
MESH geom=xyz origin=-1000 -1000 -1000 ref=0 -36.5 100
     IMESH -60 -15 -5 5 15 60 1000
     JMESH -60 -39 -36 -33 -30 -25 -20 -10 0 10 30 60 1000
     KMESH 16 30 5i 90 95 105 110 4i 160 176 1000
WWP:P 5 3 5 0 -1 j .004 j 100
DD2    -1e-4 1e3
```

The weight window mesh and values after a few iterations are shown in Figs. 4.16 and 4.17. Weight window values are very forgiving and do not have to be precise: they affect problem efficiency, not physics or answers. What is important is that the window values increase toward unimportant regions and decrease toward important regions and do not change by more than a factor of 4 between adjacent regions.

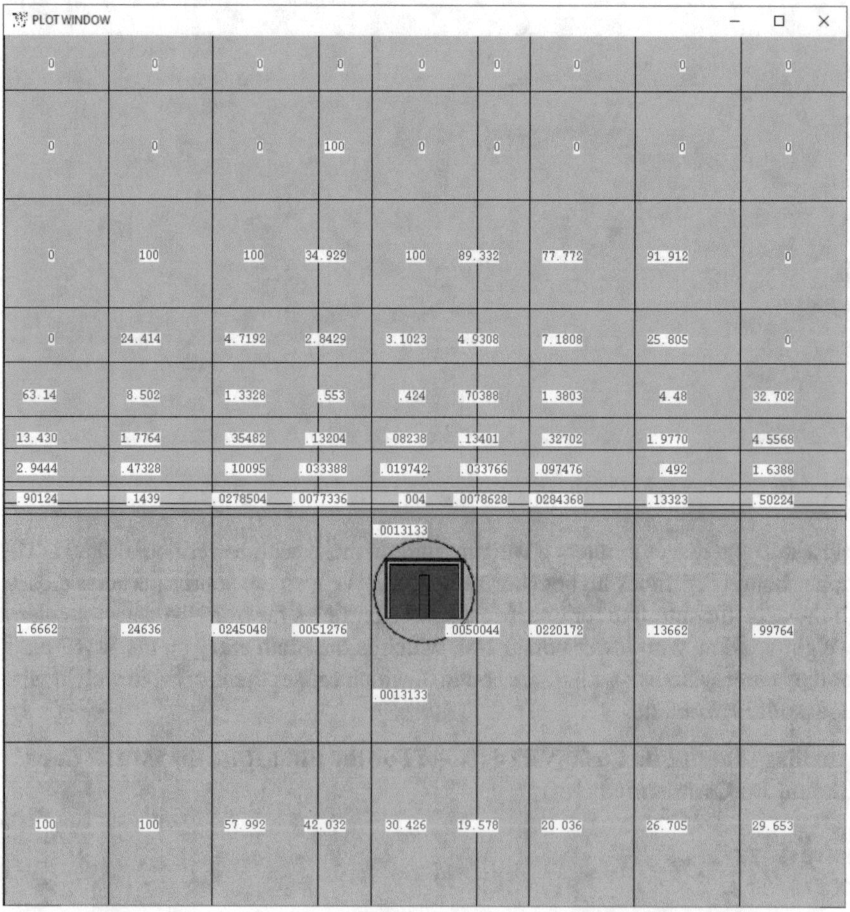

Fig. 4.17 Weight window mesh along cask axis

The plot commands are `PZ=100  OR=0  −35  100  EX=40  LA  0  1 WWN1:P MESH=3`

The plot commands are `BA=0  0  1  0  1  0  OR=0  -35  100  EX=40  LA  0 1  WWN1:P  MESH  3`

To run the problem without variance reduction techniques, simply delete the

```
READ File=Cask_VRT4
```

in the main input file, Cask4. The results are

```
          No VRT                            VRT
        Tally    error slope FOM    Tally    error slope FOM
F16  1.8756E-05 .0196 10    96  1.9124E-05 .0062 3.9 4890
F101 1.8216E-04 .0123  3.6 243  1.8354E-04 .0053 4.3 6767
```

That is, the DXTRAN/source axial bias/weight window combination is 28–50 times more efficient than the analog calculation. With the axial source position bias and no weight windows, the FOMs for tallies F16 and F101 are 3716 and 5436.

Observations
- Source axial position biasing and weight windows require only one detector be modeled at a time. The best result is with both the source axial bias and weight windows, but the weight windows are challenging to iterate and normalize and only 50% better.
- Source axial position biasing is 22–39 times more efficient than analog sampling.
- When weight windows are added, the biasing is 28–50 times more efficient than analog sampling.
- Results may have been enhanced for the case without weight windows by using importances to split particles leaving the UF_6 and entering the two layers of the cask casing.

Recommendations
- Distributed sources make weight window generation challenging.
- The weight window reference point (MESH REF) should be near the region of the desired tally region in a mesh zone, where particles are certain to pass and then score.
- The upper bound of the windows should be limited with the ninth WWP entry to prevent excessive Russian roulette.
- The seventh WWP normalization entry should be such that source particles split or are born with weights in the window. Most of the advantage of the weight windows is to eliminate by Russian roulette particles that are not of interest. The only way to determine the mesh-based weight window bounds is to plot them. (The seventh and ninth WWP entries need to be omitted or set to "J" so that the file values, not renormalized values, are plotted.) See Sect. 4.1.2.

- The weight window generator WWG

    - should be based on a tally that draws particles to the region of interest—in this case, F16—not the tally of actual interest (F18);
    - when doing iterations to set the seventh WWP entry, it is usually easier if the WWG third entry is 1 for the distributed source case. For non-distributed sources, when the source enters only into mesh zones with the same window bounds, as with a point source, the third entry is usually set between .25 and .5.

- Rectangular mesh weight windows are about 50% more efficient than cylindrical mesh windows, which are about 50% more efficient than spherical mesh windows, but is problem dependent.
- If all generated windows are zero, then no tally was scored to the WWG target tally by particles passing through the mesh that contains the SDEF REF reference point.
- If energy or time-dependent windows are used, the WWG weight at the SDEF REF point mesh will be that of the most important energy to the tally, not the most important energy group of the source. The WWP seventh entry mesh normalization must be done with care.

4.3   Neutron Detector Operation in More Detail

4.3.1   Introduction

In many applications, it is sufficiently accurate to calculate the neutron reaction rate in the detector (for example, the ^{3}He(n,p) rate in a ^{3}He detector) and use this as the estimate of the pulse-counting rate of the detector. This is a good approximation for ^{3}He and BF_3 tubes in typical usage; however, a more-detailed approach is necessary for some cases with these detectors and is essential for detectors such as boron-lined proportional counters, where a fraction of the reaction product energy is deposited in the boron layer rather than the gas.

Other cases include ^{4}He detectors and organic scintillators, in which the signal is produced by recoil of ^{4}He or H nuclei created by elastic scattering of neutrons (see Example 2.20). Note that for ^{3}He spectrometers, both ^{3}He(n,p) reactions and ^{3}He elastic scatters contribute to the measured rates. In all of these cases, a neutron reaction does not necessarily lead to a pulse above the threshold in the detector, and thus, the more-detailed treatment is necessary. MCNP6 has the capability to calculate these cases at the expense of a slightly more complicated input and longer running time. The necessary steps to include in the input file are to

1. create the reaction products and/or elastic recoils,
2. track these charged particles in the actual detector gas, and
3. tally the energy deposited in the active volume of the detector.

The procedure for carrying out these steps will now be described and followed by some examples.

4.3.1.1   Make Reaction Products (Model, Data) and Recoil Nuclei

The creation of reaction products and recoil nuclei in MCNP6 is controlled with the PHYS:N card. The following entries are allowed (MCNP manual [1] Section 3.3.3.2.1)

```
PHYS:N emax emcnf iunr J J J coilf cutn ngam J J i_int_model
i_els_model
```

The following are the default values.

```
PHYS:N 100 0 0 J J J 0 -1 1 J J 0 0
```

These default values result in the following behavior (ignoring the J placeholders):

emax = 100	The upper neutron energy limit is 100 MeV (okay but unnecessarily large for most safeguards cases);
emcnf=0	Implicit capture for all neutrons. In many cases, we want emcnf = emax or CUT:N 2J 0 0 to give analog capture, although use of the FT8 CAP or CGM options forces analog capture in any case;
iunr = 0	Unresolved resonance treatment is on;
coilf = 0	No light ion recoils are created and no reaction products are created. This setting is discussed below.
cutn=-1	When tables are available, use them to their upper limit for each nuclide, then use a physics model above that limit. Because of the important neutron energy range in most safeguards calculations, physics models are not reliable and should not be used. If the cross-section files are selected appropriately, the table data are sufficient, and this default setting is satisfactory, but by setting cutn > emax or using MPHYS OFF, all model physics can be definitively avoided (with the additional benefit that memory usage is reduced).
ngam = 1	Photons are produced using ACE. In this section, we are not tracking photons and the setting is not important.
i_int_ model = 0	Process all interactions
i_els_ model = 0	Elastic scattering by Prael/Liu/Striganov model

The MCNP code can produce light ions from neutron capture reactions in certain isotopes in certain energy ranges. The name of the model, which uses standard kinematic equations, is Neutron Capture Ion Algorithm (NCIA). This model was introduced to overcome the problem of missing particle production information in

Table 4.2 Reactions modeled by the NCIA algorithm

Nuclide (ZAID)	Reactions
^{3}He (2003)	^{3}He(n,p)t
	^{3}He(n,d)d
^{6}Li (3006)	^{6}Li(n,α)t
^{10}B (5010)	^{10}B(n,α)^{7}Li

cross-section data files. Eventually it may no longer be needed, but currently it is a reliable method to produce reaction products, as shown in Table 4.2:

The behavior we require for detailed detector calculation is to use the NCIA algorithm for reaction products (unless you are sure that the data files have reliable particle production values). You may or may not need the production of recoils, which is controlled by the parameter "coilf"; the options are given in the following table taken from the MCNP manual [1] (Section 3.3.3.2.1 Table 3-39). In summary, if coilf = 4.0, there will be one recoil per collision and NCIA will be used for reaction products; if coilf = 5, there will be no recoil production and NCIA will be used. (For coilf = 2 or 3, the table data will be used in preference to NCIA.)

Input Parameter	Description
coilf=n.m	Light-ion and heavy-ion recoil and NCIA control. (See discussion after the table.) In this format, *n* is an integer and *m* is a specified fractional value. If 0<*m*≤1 and *n*=0,1,2,or 4, then *m* is the number of light ions (protons, deuterons, tritons, ^{3}He, and alphas) per incident neutron to be created at each neutron elastic scatter event with light nuclei H, D, T, ^{3}He, and ^{4}He. Heavy ions are also created if they are specified on the MODE card. If *n*=3 or *n*=5, then *m*=0 and light-ion recoil is turned off. For *n*=2 or *n*=3, NCIA is active only when the production of NCIA ions (see table below) is not modeled with the nuclear data tables. For *n*=4 or *n*=5, NCIA is active and the nuclear data tables for production of NCIA ions are not used. Using the above set of criteria, we obtain the following description for valid *coilf* entries: If *coilf* = 0 then light-ion recoil is off; NCIA is off. (DEFAULT) If .001 < *coilf* < 1.001 then light–ion recoil makes *coilf* ions from elastic scatter. If 1.001 < *coilf* < 2.001 then light-ion recoil makes *coilf*-1 ions from elastic scatter; NCIA ions from neutron capture.* If *coilf* = 3 then light-ion recoil is off; NCIA ions from neutron capture.* If 3.001 < *coilf* < 4.001 then light-ion recoil makes *coilf*-3 ions from elastic scatter; NCIA ions from neutron capture.† If *coilf* = 5 then light-ion recoil is off; NCIA ions from neutron capture.† * Table data ion production will be used if possible † NCIA will be used even if table data are available.

4.3.1.2 Track Created Particles in Real Gas Composition

The reaction products and recoils for the nuclides shown in Table 4.2 include protons, deuterons, tritons, helions, alphas, and lithium ions. To track these particles, we need to include them on the mode card (along with neutrons). The symbols for these particles, in the same order, are H, D, T, S, A, and #.

To transport these particles correctly, it is necessary to change the lower-energy cutoff for each particle type to 0.001 MeV on the second entry of the cut:x card:

```
cut:h,d,t,s,a,#  j 0.001
```

Because we are tracking charged particles, it is essential to include all gases in the defined material description of the detector gas and not just helium. For example, most ^{3}He detectors have other gases to shorten the charge particle track length and improve the electrical properties of the gas. Many ^{3}He tubes have argon and methane (or carbon dioxide) in addition to helium.

4.3.1.3 Tally Energy Deposition of Particles in Active Volume (F8 CAP EDEP for Coincidence/Multiplicity Counting)

The F8 tally for neutron coincidence and multiplicity counting was described in Sect. 3.2.4, where the tally was used with the CAP and GATE keywords on the FT card. In the current case for charged particle tracking, we also need to use the EDEP keyword. The "EDEP tg tt" keyword causes a triggering event when the energy scored in tally tg is greater than tt (MeV). The following is an example from the MCNP manual (Chapter 3-241):

```
*F8:H,T 999
F18:N 999
FT18 CAP EDEP 8 0.001
```

A capture is scored in Tally 18 whenever there is an F8 tally that exceeds 0.001 MeV.

4.4 Examples

4.4.1 ^{3}He Detector Pulse Height

The following is an example of the calculation of the pulse-height spectrum in the ^{3}He tubes used in Example 3.2. The MCNP input file of Example 3.2 becomes Example 4.7.0 by making the following modifications:

1. The material filling the detector tubes is changed from pure ³He at a density of 4.99E–04 g/cm³ to a mixture of ³He and argon at a density of 2.208e–03 g/cm³, giving the same ³He number density:

```
10   36   -2.028e-3 -12   imp:n=1
...
c       material-36, helium-3 gas & Ar (4 atmosphere pressure =
        -2.028e-3g/cm3 at 293K)
m36   2003.      -0.24614
      18040.     -0.75386
```

2. The mode card and the tracking options were added.

```
mode n h d t s a
phys:n   25 25 0 J J J 4 30 J J J 0 0 $NCIA and light ion recoil
phys:h,d,t,s,a 20 3J 0 $ Max E=20, defaults
cut:h,d,t,s,a  j 0.001 $ lower threshold 0.001 MeV
```

3. The pulse-height tally was added.

```
c pulse height tally in 3He
f68:h,t,s (10 11 12 13 14 15 16 17 18 19 20 21 22 23 24 25 26 27)
e68   0 1e-5 10e-3 98i 2000e-3
```

Figure 4.18 shows the result of the F68 tally. This tally is the total of protons, tritons and recoil ³He nuclei (particle type s). The individual contribution could be

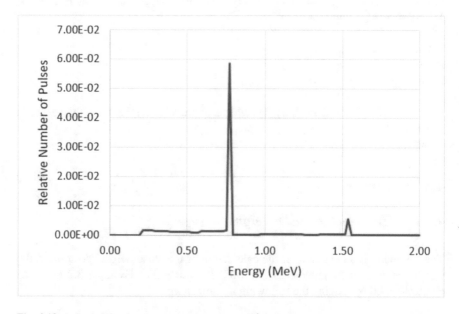

Fig. 4.18 Pulse-height spectrum inside the sum of all ³He detector tubes

extracted by separate tallies per particle type. The most prominent feature is the 762-keV peak that results from the capture of a thermal neutron. There are also the lower-energy edges in the spectrum that correspond to the wall effect on protons and tritons. Maybe more surprisingly, we see a peak at 1524 keV (2 × 762 keV). Because the source in this case is ^{252}Cf, with 3.7 neutrons per spontaneous fission, and the f8 tally records the energy deposited per history, then we have the possibility of thermal neutron captures in two different (or the same) detector tubes. Figure 4.19 shows this spectrum and one from an individual tube (on a logarithmic scale). The spectrum from the single tube shows a much smaller effect from multiple neutron events.

When a number of detector cells are combined in this F68 tally, the results are equivalent to combining the electronic pulses from all of the detectors into a single preamplifier (without the actual effect of shaping time constants, etc.).

4.4.2 ^{3}He Detector Coincidence Calculation

The operation of the EDEP tally can be shown in the simple case of the ^{3}He tube example. The input file of Example 3.2 as modified in Sect. 4.4.1 is further modified by adding the following tallies to become Example 4.7.0b:

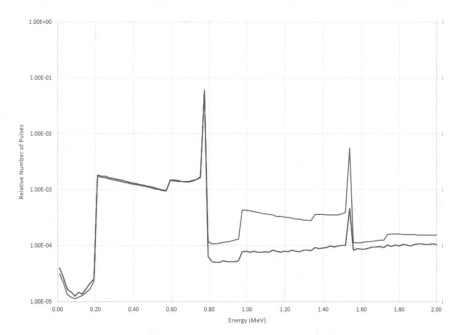

Fig. 4.19 Comparison of pulse-height spectrum in all detector tubes in blue and the pulse-height spectrum in a single tube in red. The blue curve for all tubes is from Fig. 4.18 but looks different because of the logarithmic scale instead of a linear scale

```
c tallies based on energy deposition
f6:h,t (10 11 12 13 14 15 16 17 18 19 20 21 22 23 24 25 26 27)
fc28 Ungated edep Coincidence tally
f28:n 10 $ particle type and cell doesn't matter
ft28   cap edep 6 0.001
c
fc38 Gated edep Coincidence tally
f38:n (10 11 12 13 14 15 16 17 18 19 20 21 22 23 24 25 26 27)
c    1us=10 shakes --> 4.5 us predelay and 64 us gate:
ft38   cap 2003 gate 450 6400
```

The new F6 tally records energy deposition in the sum of all ^{3}He tubes. The FT28 card of the F28 tally specifies CAP for coincidence calculations based on the result from the f6 tally with an energy threshold of 0.001 MeV. (The particle type and cell number specified on the F28 card are required but do not affect the results.) The FT38 card of the F38 tally additionally adds a coincidence predelay of 4.5 μs and a coincidence gate of 64 μs. The results from these tallies are compared in Table 4.3. Case A is the result of Example 3.2. Case B is the result of modifying Example 3.2 by the changes and additions of Sects. 4.4.1 and 4.4.2. The differences are consistent with the statistical uncertainties, giving confidence in the correct operation of the EDEP option. (*NOTE: The smaller statistical uncertainties for the new calculation arise simply because that case was run for 1E8 source events rather than 1E6 source events, as was the case for Case A, namely* Example 3.2. *Tracking the charged particles takes considerably more CPU time than tallying neutron captures.*)

4.4.3 ^{10}B-Lined Detectors

To carry out a similar calculation using boron-lined tubes instead of ^{3}He tubes, it is necessary to change the geometry used in Sect. 4.4.2 to include a layer of boron carbide around the counting gas region and to remove the ^{3}He from the counting

Table 4.3 Comparison of coincidence tallies from neutron capture (Case A, Example 3.2) and from energy deposition (Case B, Example 3.2 modified by changes in Sects. 4.4.1 and 4.4.2)

Quantity	Case A	Error	Case B	Error
	F8 (by weight)		**F28 (by weight)**	
^{3}He	2.03017E−01	0.0015	2.02997E−01	0.0002
^{3}He(^{3}He−1)/2!	4.84752E−02	0.0050	4.82156E−02	0.0005
^{3}He(^{3}He−1)(^{3}He−1)/3!	1.10729E−02	0.0185	1.07942E−02	0.0018
Quantity	**Case A**	**Error**	**Case B**	**Error**
	F18 gated (by weight)		**F38 gated (by weight)**	
n	3.43869E−02	0.0054	3.41812E−02	0.0005
$n(n − 1)/2!$	5.59098E−03	0.0250	5.45021E−03	0.0020

gas. The thickness of the boron carbide layer makes a large difference to the results, so different cases (0.7, 2, and 4 μm) will be shown.

Example 4.7 is for the 2 μm thick case. The boron-lined layer thickness is varied by adjusting the radii of surfaces 12 and 13. The different thresholds are tallied with different EDEP thresholds in tallies F228, F238, F248, F258, and F268.

Example 4.7 MCNP Input File for the 2 μm Thick, Boron-Lined Layer Case (90% ^{10}B, 2.52 g/cm^3)

```
Crude model of HLNC2 based on p499 of PANDA Manual
c    Cells
c    sample volume
100 94 -2.5 -100 imp:n=1
C    Cavity
1    0 -1 100 imp:n=1
C    Bottom end plug poly
2    1 -0.96 -2 3  imp:n=1
C    Bottom end plug Al
3    13 -2.7 -3  imp:n=1
C    Bottom end plug Cd
4    48 -8.64 -4 2 imp:n=1
C    Top end plug poly
5    1 -0.96 -5 6  imp:n=1
C    Top end plug Al
6    13 -2.7 -6 imp:n=1
C    Top end plug Cd
7    48 -8.64 -7 5 6  imp:n=1
c    Cd liner
8    48 -8.64 -11 1 4 7 imp:n=1
c    Poly moderator with external cutouts and holes for #He tubes
9    1 -0.96 (-8  : -9  : -10 ) 11 12 #11 #12 #13 #14 #15 #16 #17 #18
                        #19 #20 #21 #22 #23 #24 #25 #26 #27
                        #31 #32 #33 #34 #35 #36 #37 #38
                        #39 #40 #41 #42 #43 #44 #45 #46 #47
                        imp:n=1
C    18 counting gas volumes (no walls, no air gap, no top and bot-
tom end spaces)
10   43  -1.664e-3 -13  imp:n=1
11   Like 10 but TRCL=(-0.730      4.138  0 )
12   Like 10 but TRCL=(-2.831      7.778  0 )
13   Like 10 but TRCL=(-6.050     10.479  0 )
14   Like 10 but TRCL=(-9.999     11.916  0 )
15   Like 10 but TRCL=(-14.201    11.916  0 )
16   Like 10 but TRCL=(-18.150    10.479  0 )
17   Like 10 but TRCL=(-21.369     7.778  0 )
```

```
18   Like 10 but TRCL=(-23.470      4.138    0 )
19   Like 10 but TRCL=(-24.200      0.000    0 )
20   Like 10 but TRCL=(-23.470     -4.138    0 )
21   Like 10 but TRCL=(-21.369     -7.778    0 )
22   Like 10 but TRCL=(-18.150    -10.479    0 )
23   Like 10 but TRCL=(-14.201    -11.916    0 )
24   Like 10 but TRCL=(-9.999     -11.916    0 )
25   Like 10 but TRCL=(-6.050     -10.479    0 )
26   Like 10 but TRCL=(-2.831      -7.778    0 )
27   Like 10 but TRCL=(-0.730      -4.138    0 )
c
c   Boron layers
c
30   42   -2.52 -12 13  imp:n=1
31   Like 30 but TRCL=(-0.730       4.138    0 )
32   Like 30 but TRCL=(-2.831       7.778    0 )
33   Like 30 but TRCL=(-6.050      10.479    0 )
34   Like 30 but TRCL=(-9.999      11.916    0 )
35   Like 30 but TRCL=(-14.201     11.916    0 )
36   Like 30 but TRCL=(-18.150     10.479    0 )
37   Like 30 but TRCL=(-21.369      7.778    0 )
38   Like 30 but TRCL=(-23.470      4.138    0 )
39   Like 30 but TRCL=(-24.200      0.000    0 )
40   Like 30 but TRCL=(-23.470     -4.138    0 )
41   Like 30 but TRCL=(-21.369     -7.778    0 )
42   Like 30 but TRCL=(-18.150    -10.479    0 )
43   Like 30 but TRCL=(-14.201    -11.916    0 )
44   Like 30 but TRCL=(-9.999     -11.916    0 )
45   Like 30 but TRCL=(-6.050     -10.479    0 )
46   Like 30 but TRCL=(-2.831      -7.778    0 )
47   Like 30 but TRCL=(-0.730      -4.138    0 )
c    Outside vacuum
110   0   -999 8 9 10 imp:n=1
c    Exterior world
120   0 999 imp:n=0

c    surfaces
1    rcc 0     0 -20.3412 0 0 40.6824 8.75    $ cavity
2    rcc 0     0 -28.8     0 0 8.3     8.75    $ Bottom end plug poly
3    rcc 0     0 -26.7     0 0 6.2     6.75    $ Bottom end plug Al
4    rcc 0     0 -28.8     0 0 8.4588  8.75    $ Bottom end plug Cd
5    rcc 0     0  20.5     0 0 8.3     8.75    $ Top end plug poly
6    rcc 0     0  20.5     0 0 6.2     6.75    $ Top end plug Al
7    rcc 0     0  20.3412  0 0 8.4588  8.75    $ Top end plug Cd
8    rcc 0     0 -28.8     0 0 16.8    16.9    $ bottom poly moderator
```

```
9    rcc 0     0 -12.      0 0 24      15     $ middle poly moderator
10   rcc 0     0  12       0 0 16.8    16.9   $ top poly moderator
11   rcc 0     0 -28.8     0 0 57.6    8.9088 $ Cd liner
12   rcc 12.1 0 -25.4      0 0 50.8    1.27   $ Boron layer   (B4C)
13   rcc 12.1 0 -25.4      0 0 50.8    1.2698 $ counting gas
100  rcc 0     0 -10       0 0 20      5      $ Sample volume
999  rcc 0     0 -30       0 0 60      20


C    Materials
c    poly density 0.96 g/cm3
m1   1001 2 6012 1
mt1  poly.20t  $ specification of thermal S(a,b) scattering
c    Al
m13 13027 1
c    Cd
m48 48000 1
c    material-42 boron10 carbide (90% 10B)
   m42  5010.    0.72
        5011.    0.08
        6000.    0.20
c    material-43, Ar 1atm 293K = 1.664e-3 g/cm3
   m43  18040.70c    1.0
c plutonium oxide
m94  94240 0.2 94239 0.8 8016 2.0
fmult   94239 method=5 data=3 shift=1 $ method=LLNL data=Ensslin etc
c                                      shift=MCNPX
fmult      94240 method=5 data=3 shift=1 $
C
c
sdef par=sf  ext=d1 axs = 0 0 1 rad=d2
si1    -10 10
sp1    0    1
si2    0    5
sp2    -21 1
c
c
print     -85 -86
nps 1e8
c  neutron alpha 7Li
mode n a #
phys:n,a,#    25 25 0 J J J 5 30 J J J 0 0 $ NCIA and no light
ion recoil
cut:a,#   j 0.001 $ lower threshold 0.001 MeV
c
c tallies (FOR 10B )
```

```
c pulse height tally in counting gas
f68:a,# (10 11 12 13 14 15 16 17 18 19 20 21 22 23 24 25 26 27)
e68   0 1e-5 10e-3 248i 2500e-3
c reactions in boron layer cells
f4:n (30 31 32 33 34 35 36 37 38 39 40 41 42 43 44 45 46 47)
fc4 Number of 10B(n,a) reactions
fm4 -1 42 107
sd4 1
fc58 Ungated Coincidence tally in boron layers (not detectable)
f58:n (30 31 32 33 34 35 36 37 38 39 40 41 42 43 44 45 46 47)
ft58   cap 5010
c energy deposition tally for coinc edep tally
f6:a,# (10 11 12 13 14 15 16 17 18 19 20 21 22 23 24 25 26 27)   $
counting gas
fc8 Edep of alpha and 7Li
f8:a (10 11 12 13 14 15 16 17 18 19 20 21 22 23 24 25 26 27)   $
particle type and cell number are irrelevant
ft8 cap edep 6 0.001
c
fc228 Zero energy threshold
f228:n 6
ft228 cap edep 6 0
f238:n 6
fc238 0.15 Mev energy threshold
ft238 cap edep 6 .15
f248:n 6
fc248 0.5 MeV energy threshold
ft248 cap edep 6 .5
f258:n 6
fc258 0.7 MeV energy threshold
ft258 cap edep 6 .7
f268:n 6
fc268 0.9 MeV energy threshold
ft268 cap edep 6 .9
```

For this calculation, we need to track α particles and ^{7}Li reaction products on the mode card

```
mode n a #
```

and set the appropriate tracking options for these particles.

```
phys:n,a,#    25 25 0 J J J 5 30 J J J 0 0 $ NCIA and no light
ion recoil
cut:a,#   j 0.001 $ lower threshold 0.001 MeV
```

We can calculate the pulse height in the same way as the F68 tally for ^{3}He except using a and # particle types. The results are shown in Fig. 4.20.

The results from the tallies for the various layer thicknesses are shown in Table 4.4 and will now be explained. The "number of ^{10}B reactions" comes from the production of α-particles (and ^{7}Li) given in the accounting table, which agrees with the F4 tally of (n,α) reactions in the boron carbide layers and with the F8 CAP tally

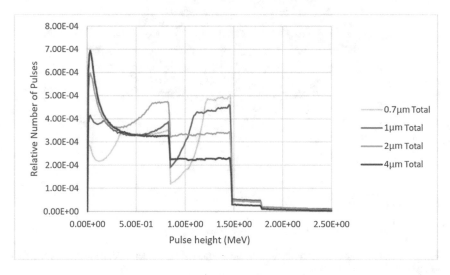

Fig. 4.20 Pulse height in boron-lined detectors of different layer thicknesses (0.7, 1, 2, and 4 μm)

Table 4.4 Results from the sum of all boron-lined detector tubes

Layer thickness					
	Threshold MeV	0.7 μm	1 μm	2 μm	4 μm
Number of ^{10}B reactions		6.26E−02	7.89E−02	1.17E−01	1.56E−01
F4		6.22E−02	7.90E−02	1.17E−01	1.56E−01
F8 in B$_4$C		6.22E−02	7.89E−02	1.17E−01	1.56E−01
Single events in gas >threshold	0	5.21E−02	6.03E−02	6.38E−02	5.04E−02
	0.15	4.80E−02	5.36E−02	5.51E−02	4.10E−02
	0.5	3.63E−02	3.99E−02	4.03E−02	2.73E−02
	0.7	2.90E−02	3.24E−02	3.06E−02	2.02E−02
	0.9	2.30E−02	2.54E−02	2.11E−02	1.39E−02
Doubles events in gas >threshold	0	1.57E−03		2.76E−03	
	0.15	1.34E−03		2.06E−03	
	0.5	7.65E−04		1.10E−03	
	0.7	4.85E−04		6.33E−04	
	0.9	3.06E−04		3.04E−04	

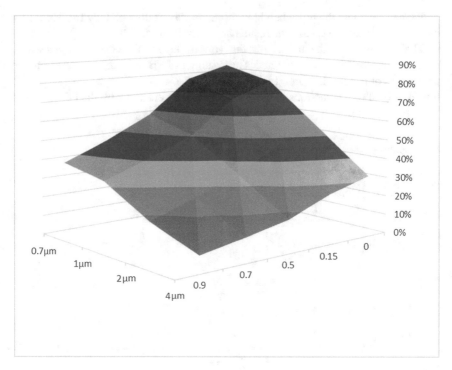

Fig. 4.21 Percentage of neutron captured in the boron layer that is detected in the gas as a function of layer thickness and threshold

in the same layers. (*Note: This quantity cannot normally be measured experimentally.*)

The Doubles events in Table 4.4 come from the first entry in the various gated F8 tallies. These values were calculated for only the 0.7- and the 2-μm cases.

The "single events in the gas" greater than various thresholds come from the first entry (by weight) of the various ungated F8 tallies based on the F6 energy-deposition tally. These entries are shown in Fig. 4.21. The maximum of 83% is for a 0 MeV threshold and a 0.7 μm thick B_4C layer; the minimum is 9% for a 0.9 MeV threshold and a 4 μm thick layer. These values were referred to as the "electronic efficiency" in [4]. If we assume that we need a detection threshold of at least 0.15 MeV, then the maximum efficiency of the full boron-lined system occurs at a layer thickness of 2 μm and is 5.5%, compared with the 20.3% efficiency of the ^{3}He-based system, which uses 4 atm pressure (and so could be further improved).

Using gated tallies with different gate widths, we can calculate the dieaway time of the detector, as shown in Fig. 4.22. The dieaway time is 69.6 μs compared with the ^{3}He HLNCC2 detector's 43 μs, which would also worsen its relative performance. (The optimum gate width of this detector is closer to 90 μs than the 64 μs used in these calculations.)

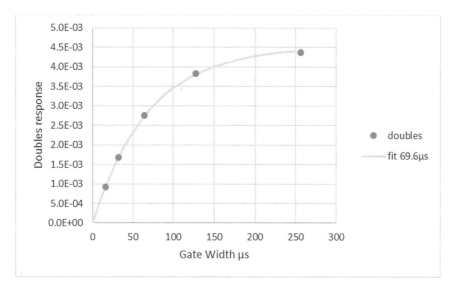

Fig. 4.22 Doubles response vs. gate width and a fit to the data with a dieaway time of 69.6 μs

References

1. C.J. Werner, *MCNP User's Manual Code Version 6.2*, LA-UR-17-29981 (2017)
2. Oak Ridge National Laboratory (n.d.). https://www.ornl.gov/onramp/advantg
3. X-5 Monte Carlo Team, *MCNP—A General Monte Carlo N-Particle Transport Code, Version 5, Volume I: Overview and Theory*, Los Alamos National Laboratory report LA-UR-03-1987 (2003), Revised February 1, 2008
4. M. T. Swinhoe, J. S. Hendricks, Calculation of the performance of ^{3}He alternative detectors with MCNPX, in *2011 2nd International Conference on Advancements in Nuclear Instrumentation, Measurement Methods and their Applications (ANIMMA 2011)*, Ghent, Belgium (June 6–9, 2011), pp. 1–6

Chapter 5
Additional Topics

5.1 Troubleshooting or "How Can I Be Confident in the Results?"

This chapter is concerned with the question "Is the Monte Carlo calculation that I just carried out 'correct' or adequate for the intended purpose?"

Unfortunately, there are many more ways that a calculation can be wrong than right and the input and output need to be carefully checked before the results can be relied upon. We will consider many factors that can affect the results and we will concentrate on neutron examples. We will be considering modeling errors, i.e., mistakes. (Statistical uncertainties are considered in Sect. 2.6.)

The principal sources of errors are

- Geometry and materials
- Detector modeling
- Source modeling
- Sample modeling
- Tracking limitations
- Nuclear data
- Statistics
- User

These errors will be discussed in the sections that follow. The best approach, where possible, is to carry out a benchmarking experiment. Working on both experiments and modeling of a measurement situation usually results in improvement in both. Of course, such benchmarking is not always possible, in which case, simulation models may be checked with a simpler case, for example, the use of ^{252}Cf instead of plutonium. Then the basic validity of the model is checked, eliminating many potential errors in the complete simulation.

Supplementary Information The online version contains supplementary material available at [https://doi.org/10.1007/978-3-031-04129-7_5].

J. S. Hendricks et al., *Monte Carlo N-Particle Simulations for Nuclear Detection and Safeguards*, https://doi.org/10.1007/978-3-031-04129-7_5

A best practice guide was prepared by the ESARDA NDA working group [1], which contains many practical recommendations. This guide distinguishes between soft cases, where the required uncertainty in the result is around 10%, and hard cases, where the desired uncertainty is around 1%.

5.1.1   Geometry and Materials

It is impossible to make an exact model of a real-life situation. The details are too complex and many parameters are not well known. Fortunately, many of these details are unimportant and have a negligible effect on the result (as well as complicating the geometry and slowing down the calculation). It is up to the user to differentiate between important and unimportant features. In many cases, the exact dimensions of parts of detectors are not well known—internal spaces that depend on manufacturing tolerance, for example.

Some commonly used materials may have unknown density in the as-built equipment. In particular, the density of "high-density" polyethylene has some variations. The Pacific Northwest National Laboratory guide to composition data for radiation transport modeling [2], which is a very useful guide for materials composition, gives the density of "high-density" polyethylene (non-borated)—often used in nuclear instrumentation—as ranging between 0.944 and 0.965 g/cm^3, whereas in our experience, the density is closer to the upper end of this range. If the polyethylene density (which can be verified with scoping calculations) is important, then the actual density should be measured.

Another issue is the presence of unknown impurities. Most impurities occur at such low levels that they do not significantly affect neutron transport; however, one example known to the authors was that some lead used for gamma shielding in a neutron detector contained sufficient cadmium to reduce the measured detection efficiency by 30%. This example was discovered only because a second instrument with nominally identical construction had the expected efficiency. The detector environment (other equipment, walls, floors, etc.) may also have a significant effect, which depends on the detector itself. Large, heavily shielded well counters are generally less sensitive to many of these effects.

5.1.2   Detector Modeling

There are many important factors in modeling ^{3}He detectors, for example. The tube dimensions are obviously important, as well as the amount of ^{3}He. Other high-Z gases that are present are important if the reaction products from ^{3}He(n,pt) are to be tracked. The presence of end "dead" zones—filled with ^{3}He but where the deposited charge is not collected—nevertheless affects the counting rate in the active length of the tube. The main factor is the absorption of thermal neutrons that otherwise could have been captured in the active volume. A much smaller effect is reaction product, created in the dead zones, that enters the active volume. In many cases, it is sufficient to use the ^{3}He(n,pt) rate in the detector active volume as a measure of count rate, but there are several important steps between the occurrence of a reaction and the production of an electronic pulse that is recorded. These steps include overcoming electronic thresholds and deadtime. The effect of the threshold, which typically has a less than 1% effect on ^{3}He detector systems, is vitally important when modeling ^{4}He- or ^{10}B-lined detectors. For this kind of detector, the F8 tally can be used with the EDEP keyword to apply a threshold in energy to record only those events that deposit sufficient energy to create a pulse. Another real-life effect is deadtime, which is not treated here. A deadtime correction of experimental data is normally needed before experimental results are compared with simulation results.

5.1.3 Source Modeling

Modeling of the source can have important effects on the result of the simulation. The absolute source strength is a parameter that does not enter into the running of the simulation itself but directly affects the answer. Many neutron sources (for example, ^{252}Cf and AmLi) in common use do not have accurate values for their absolute source strength. Many manufacturers provide nominal values for source strength that can be as far as 15–20% from the true value, which is one of the most important issues regarding getting agreement between simulated and experimental results.

One of the most important factors within the simulation itself is the neutron source spectrum. The importance of the spectrum depends on the configuration being modeled; quantities can be relatively sensitive to the spectrum, but each case should be carefully considered. For example, the uncertainty in the mean energy of spontaneous fission neutrons from ^{240}Pu was estimated to be 50 keV [3]. A change of 50 keV to the mean energy gave a 0.4% change in efficiency of an epithermal neutron multiplicity counter, a detector less sensitive than many others to neutron spectrum.

(Alpha,n) neutron sources are more challenging to model, and the case of AmLi sources is of particular importance to safeguards. Such sources are used because they emit single neutrons (not coincidence neutrons like spontaneous fission sources) and the average neutron energy is low so that the neutrons have a low probability of causing fission in ^{238}U. Good investigations of the AmLi neutron spectrum and its effect on measurements are given in [4–6]. These papers have collected AmLi spectra from different origins and compare the results of many simulations that used different spectra. An important distinction arises between a spectrum that represents the neutrons at their birth and a spectrum (usually measured) as it escapes from the container. Use of the former type in a simulation should include the transport of neutrons through its source container and material, whereas use of the latter type should not.

Even the location and extent of sources are not always obvious. A typical ^{252}Cf source capsule may be 25 mm long, but the source may be located as a small piece of wire, well away from the center of the capsule.

5.1.4 Sample Modeling

In theoretical studies, it is simple to specify the details of the item; everything is well defined. In modeling of actual experiments, it can be very difficult. A well-known case is the effect of powder density on the neutron multiplication in plutonium oxide or mixed uranium-plutonium (MOX) powder. When the material is sealed in a can, the mass can be well known but, without particular efforts such as radiography, it can be difficult to be certain of the density. It can be even more difficult to know the moisture content, the main chemical form, and the presence of impurities—all of which can significantly change neutron-counting rates. In pure, dry plutonium dioxide, there are roughly equal neutron emissions from spontaneous fission and (α,n) reactions on oxygen, but every extra 100 ppm of fluorine will make an additional 1% contribution to the neutron output.

5.1.5 Tracking Limitations

In neutron applications, the tracking algorithms of the code can be considered to introduce less error than from any other source. There may be special cases where neutron behavior in the resonance region is important, but this does not affect the majority of safeguards calculations. Although not a tracking limitation, it is important to remember to specify datasets to cover the $S(\alpha,\beta)$ treatment for thermal neutrons, for example, poly.10t for neutrons in polyethylene.

5.1.6 Nuclear Data

The current status of nuclear data for use with the MCNP code is a vast topic and we can touch only a few highlights here. For cross-sections, it is essential to check in the output file which particular cross-section dataset has been selected for each nuclide. Some resources are available to check what data is available (Refs. [7, 8]). Sometimes the specification of natural elemental data (e.g., 48000) will force the code to use old data, whereas in some cases the specification of nuclide-specific data (e.g., 6012) will have the same effect! A very important point for safeguards calculations, which generally use the neutron energy region below about 20 MeV, is that cross sections calculated using models are normally not nearly accurate enough for practical use. To ensure that models are not used, CUTN on the PHYS:N card should be set greater than EMAX, which also saves memory.

The MCNP code has a built-in set of spontaneous fission data for most nuclides of importance, shown in Fig. 5.1.

These data cover the neutron energy spectrum and the $P(\nu)$ distribution for spontaneous fission for these nuclei. They can be overridden by the user. The data for induced fission can be taken from the cross-section data files, or alternatively, the FREYA or CGMF models (MCNP manual [1] Section 3.3.3.8) can be used. The data in the cross-section files are limited on the multiplicity, energy spectrum, and angular distribution (and correlations between these quantities) of neutrons from both spontaneous and induced fission. The tabulated data files do not usually contain sufficient information, especially for the correlations, whereas the use of fission modeling can include all coupled values but sometimes at the expense of not matching well-known measured quantities, such as mean neutron energy. For example, Ref. [9] states that calculations that use data from standard cross-section files include no angular correlation, but using the FREYA fission model gives mean energy for ^{252}Cf spontaneous fission neutrons that is 150 keV too high. In addition, it is often the case that measured data are inadequate to judge the performance of the models. See [9] for details.

5.1.7   Statistics

The statistics of the result is not generally an error and is dealt with in Sect. 2.6. It is the easiest to quantify because of the large amount of analysis built into the code. There can be some problem cases where a tally does not converge (code messages should be carefully examined), in which cases it may be necessary to apply variance reduction techniques (see Sect. 4.1).

5.1.8   User

In complex geometries, the geometry can be wrong. This is a legal description of geometry (not flagged by the MCNP code with error messages or lost particle warnings). All geometries, especially complex ones, should be checked by extensive geometry plotting.

Two frequent mistakes committed by the user who is trying to reduce the running time of the problem do produce MCNP warning messages (and this is another opportunity to mention that all MCNP error messages should be examined carefully and understood before the results of a simulation are used). These two errors are the incorrect use of energy cutoffs or time cutoffs. One simple example of such an error is when only fast neutrons are of interest and an error cutoff—say 10 keV—is used; then thermalized neutrons are excluded from causing thermal fission and so

```
1fission multiplicity data.                                                   print table 38

 Gaussian widths from Ensslin, Santi, Beddingfield, Mayo (1998-2005)
 Gaussian shift by MCNPX option = 1
 Gaussian sampling by MCNPX method.
 Gaussian sampling uses xnu + rang.
 Fission multiplicity tracks MCNPX.

    zaid      width      watt1      watt2      yield    sfnu

   90232  1.079000  0.800000  4.000000  6.00E-08  2.140
   92232  1.079000  0.892204  3.722780  1.30E+00  1.710
   92233  1.041000  0.854803  4.032100  8.60E-04  1.760
   92234  1.079000  0.771241  4.924490  5.02E-03  1.810
   92235  1.072000  0.774713  4.852310  2.99E-04  1.860
   92236  1.079000  0.735166  5.357460  5.49E-03  1.910
   92238  1.230000  0.648318  6.810570  1.36E-02  0.048  0.297  0.722  0.950  0.993  1.000  1.000  1.000  1.000  1.000
   93237  1.079000  0.833438  4.241470  1.14E-04  2.050
   94236  1.079000  0.000000  0.000000  0.00E+00  0.080  0.293  0.670  0.905  0.980  1.000  1.000  1.000  1.000  1.000
   94238  1.115000  0.847833  4.169330  2.59E+03  0.056  0.267  0.647  0.869  0.974  1.000  1.000  1.000  1.000  1.000
 * 94239  1.140000  0.885247  3.802690  2.18E-02  2.160
 * 94240  1.109000  0.794930  4.689270  1.02E+03  0.063  0.295  0.628  0.881  0.980  0.998  1.000  1.000  1.000  1.000
   94241  1.079000  0.842472  4.151500  5.00E-02  2.250
   94242  1.069000  0.819150  4.366680  1.72E+03  0.068  0.297  0.631  0.879  0.979  0.997  1.000  1.000  1.000  1.000
   95241  1.079000  0.933020  3.461950  1.18E+00  3.220
   96242  1.053000  0.887353  3.891760  2.10E+07  0.021  0.168  0.495  0.822  0.959  0.996  0.999  1.000  1.000  1.000
   96244  1.036000  0.902523  3.720330  1.08E+07  0.015  0.131  0.431  0.764  0.948  0.991  1.000  1.000  1.000  1.000
   96246  1.079000  0.000000  0.000000  0.00E+00  0.015  0.091  0.354  0.699  0.917  0.993  1.000  1.000  1.000  1.000
   96248  1.079000  0.000000  0.000000  0.00E+00  0.007  0.066  0.287  0.638  0.892  0.982  0.998  1.000  1.000  1.000
   97249  1.079000  0.891281  3.794050  1.00E+05  3.400
   98246  1.079000  0.000000  0.000000  0.00E+00  0.001  0.114  0.349  0.623  0.844  0.970  1.000  1.000  1.000  1.000
   98250  1.079000  0.000000  0.000000  0.00E+00  0.004  0.040  0.208  0.502  0.801  0.946  0.993  0.997  1.000  1.000
   98252  1.207000  1.180000  1.034190  2.34E+12  0.002  0.028  0.153  0.427  0.733  0.918  0.984  0.998  1.000  1.000
   98254  1.079000  0.000000  0.000000  0.00E+00  0.000  0.019  0.132  0.396  0.714  0.908  0.983  0.998  1.000  1.000
  100257  1.079000  0.000000  0.000000  0.00E+00  0.021  0.073  0.190  0.390  0.652  0.853  0.959  0.993  1.000  1.000
  102252  1.079000  0.000000  0.000000  0.00E+00  0.057  0.115  0.207  0.351  0.534  0.717  0.863  0.959  0.997  1.000
```

Fig. 5.1 Fission multiplicity data

potential fast neutrons are missing from the simulation. Time cutoffs can be a similar problem, especially if the contribution of delayed neutrons to the problem is intended to be included.

A final concern, which is not really related to a user mistake, is worth mentioning. In the "olden days," to overcome running time problems and lack of nuclear data values in the code, some simulations were performed in two parts—often active neutron interrogation scenarios. One example would be the simulation of a Cf shuffler in which the induced fission rate (e.g., in ^{235}U) caused by the Cf source was calculated in the first part, and in the second part, delayed neutrons would be started in the item to calculate the detection efficiency. This example allows the user to choose the yield and energy spectrum of the delayed neutrons when this information was not available in the code or cross-section datasets. The potential error here is that the spatial distribution of the delayed neutrons in the second run may not properly reflect the locations of fissions caused in the first part of the simulation, thereby producing incorrect counting rates. Self-shielding in the item would be an area that could create such a problem. A similar concern exists with active neutron collar or differential dieaway simulations.

5.1.9 Checking Your Results

Always perform a "sanity check" on your results. Are they reasonable and of the correct magnitude?

Spend some time plotting the geometry to make sure it is what you intended.

Read carefully and understand all warning messages generated by the code.

If possible, compare your simulated result with the experiment. If the full experiment is not possible, test your results with, for example, a Cf source if a real sample is not available.

Use previously validated components, such as ^{3}He detectors, with all of their properties that have been used in previous models.

Carry out sensitivity scoping calculations to check the effect of variations on your final result. In any case, these calculations will be needed to make a comprehensive estimate of the overall uncertainty on your calculated result. It is relatively simple to check the effect of dimensions, densities, and source spectra. Calculations can be done with different selected cross-section datasets and different code options (unresolved resonance behavior, fission models, etc.) to check on sensitivity to those. The use of the PERT card (MCNP manual [1] Section 3.3.5.21) can be considered to make convenient changes to material density, composition, or reaction cross-section data.

5.2   Sampling Collision Progeny

The consequences of any collision modeled in the MCNP code can be sampled and plotted using the LCA card eighth entry, NOACT (MCNP manual [1] Section 3.3.3.7.2). If NOACT = −2, source particles immediately collide and all progeny escape. All secondary particles produced are transported with no interactions and no decay. With the appropriate tallies, this may be used to compute and tally

- secondary energy distributions with F1,
- secondary angle distributions with F1 and FT1 FRV,
- double-differential cross sections,
- residual nuclei with F1 or F8 with FT RES option,
- secondary particle production,
- delayed neutron production,
- comparison of model to table physics, and
- comparison of nearly any physical process in the MCNP code.

The LCA card was intended to provide physics options for the LAHET code system, but the NOACT eighth entry proved so useful that it was extended to all MCNP physics regimes.

5.2.1   Analysis of Delayed Neutron Production

Example 5.1 is an input file used to analyze the delayed neutron energy and time spectrum from a Frohner Watt spectrum incident neutron. This delayed neutron spectrum is the time and energy source of the Cf shuffler problem in Sect. 4.1.5.

Example 5.1 Input to Model ^{235}U Delayed Neutron Production from Watt Spectrum Neutron Collisions

```
Delayed Neutron Spectra and Time

1 104 -1.0 -11  imp:n=1
3   0       11  imp:n=0

11 sph 0 0 0  1.0

LCA  7J  -2
ACT  DNBIAS=15
cut:n 2J 0 0
nps 10000000
prdmp 2j 1 3
m104 92235.70c 1
sdef  erg=d3
```

```
sp3    -3 1.175 1.0401  $ Frohner Watt parameters
c
FC11 235U Frohner Watt delayed neutron energy spectrum
F11:n 11
E11   1e-9 200iLOG 20
T11   1e4 1e8 10e10 1e20 T
FQ11 E T
c
FC21 235U Frohner Watt delayed neutron time spectrum
F21:n 11
T21   1e8 8i 10e8 98i 1000e8
FQ21  T F
```

Figures 5.2 and 5.3 compare the results to the MCNP standard Watt spectrum three ways

```
sp3    -3 1.175 1.0401  $ Frohner Watt parameters
SP3    -3               $ Default Watt spectrum
```

and the default Watt spectrum is compared with two different ^{235}U data files

```
m104 92235.70c 1        $ ENDF/B-VII
m104 92235.80c 1        $ ENDF/B-VII.1
```

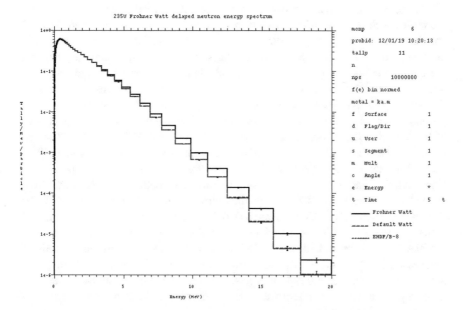

Fig. 5.2 Comparison of outgoing neutron energy spectra from Watt fission source neutrons having a single collision with ^{235}U

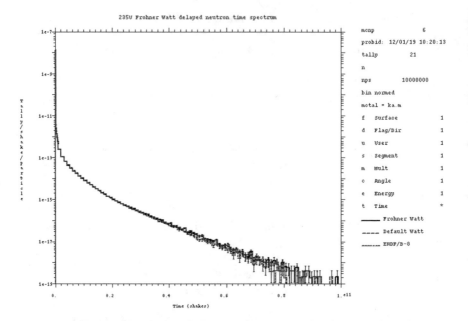

Fig. 5.3 Comparison of outgoing neutron time spectra from Watt fission source neutrons having a single collision with ^{235}U

Results are within statistical agreement regardless of whether delayed neutron bias is invoked, DNBIAS = 15. The energy spectra are unchanged with and without DNBIAS. But all delayed neutron time results, F21, tallied after 1e8 shakes (1s), converge much better with DNBIAS turned on. Relative errors range from 16 times smaller at time = 2 s to being infinitely better when after time >300 s, there are no neutrons produced with NPS = 1e8. Comparison of the problem summary tables in the output for both cases, with and without DNBIAS, shows how the physics is unchanged but DNBIAS improves the sampling of delayed neutrons. In both cases, there are 0.163 fissions per source and hence, per collision. These produce 0.437 prompt neutrons and 0.00262 delayed neutrons. The physics is within statistical uncertainty, but the sampling is very different. For no DNBIAS with 1e8 source neutrons, there are 0.163 fissions, 0.437 prompt fission neutrons, and 0.00262 delayed fission neutrons. With DNBIAS, there are 0.163 fissions, 0.437 prompt fission neutrons, and 1.36 delayed fission neutrons. In both cases, there are 2.674 prompt fission tracks per fission, but the delayed fission track production is very different. Without DNBIAS, there are 0.016 delayed neutron tracks per fission. With DNBIAS, there are 8.326 delayed fission tracks per fission. Note that, although DNBIAS = 15 is requested, the MCNP code allows a maximum bias of only 10 delayed fission neutrons per fission. Furthermore, the MCNP delayed neutron bias algorithm has to allocate the prompt fission neutrons before sampling the delayed neutron production and, consequently, only an average of 8.326 biased delayed neutrons are produced per fission.

Figures 5.2 and 5.3 show that, for the default Watt spectrum, there is no statistical difference between the 2007 evaluation of nuclide 92235.70c (ENDF/B-VII) and the 2012 evaluation of nuclide 92235.80c (ENDF/B-VII.1) results. The 92235.80c interactions (label ENDF/B-8 in the plot) have an 11% lower FOM which means 92235.70c is faster by 11%. In general, the more modern evaluations are slower but have more accurate secondary particle information.

Figures 5.2 and 5.3 demonstrate that the secondary energy and time spectra may be calculated and compared with different Watt neutron source spectra and different table cross-section libraries. The example is for delayed neutron production but applies to any collision physics. The example is for energy time and energy spectra but applies to any tally dimension, including multipliers to get reaction rates, special user tallies to get residuals, and angular distributions.

5.2.2   Comparison of Table Physics vs. Model Physics

Example 5.2 is an input file that uses the LCA 7J -2 (NOACT = -2) capability to compare table physics with model physics.

Example 5.2 Input to Compare Secondary Energy Distribution, Secondary Angle Distribution, and Secondary Particle Production from the Collisions of 1-, 15-, 100-, and 150-MeV Neutrons on ^{1}H, ^{12}C, and ^{16}O

```
Comparison of table vs model secondary energies and angles
1 101 -1.0 -11    imp:n=1
2 102 -1.0 -12    imp:n=1
3 103 -1.0 -13    imp:n=1
4   0  11 12 13   imp:n=0

11 sph 1 0 0   .5
12 sph 0 1 0   .5
13 sph 0 0 1   .5

MODE N H D T S A
LCA   7J  -2
PHYS:N 150
cut:n 2J 0 0
nps 10000000
prdmp 2j 1 3
print -30 -162 -85 -86
m101     1001 1
m102     6012 1
mx102:n 6000
m103     8016 1
```

```
SDEF  erg=d3 vec=0 0 1 dir=1 pos=d5
SI3 L 1 15 100 149.9
SP3   1  1   1   1
SI5 L  1 0 0   0 1 0    0 0 1
SP5    1       1        1
c
FC11 Neutron energy spectrum
F11:n 11 12 13
E11   1e-9 200iLOG 150
FT11  SCX 3
FQ11 F E U
c
FC21 Angular distribution
F21:n 11 12 13
C21  -.99 198I 1.0 T
FT21  FRV 0 0 1  SCX 3
FQ21 F C U
```

To change the input from table to model physics, the materials cards are replaced as:

```
m101     1001 1
mx101:n model
m102     6012 1
mx102:n model
m103     8016 1
mx103:n model
```

Cells 1, 2, and 3 are sampled equally and independently to compare neutrons incident on ^{1}H, ^{12}C, and ^{16}O. For each material, the incident neutrons are at four energies: 1, 15, 100, and 150 MeV. The F11 tally estimates the secondary energy distribution for each of these 12 combinations. The F21 tally estimates the secondary angular distribution relative to the incident VEC = 0 0 1 DIR = 1 incident source direction for each of these 12 combinations. The FT SCX option enables the contributions from each of the four incident energies to be tallied separately.

Secondary particle production is invoked by

```
MODE N H D T S A
```

The summary tables in the output file of Example 5.2 provide the secondary particle production as the source weight for each particle type. Table 5.1 shows the physical number (weight) of each particle type per source neutron (weight = 1.0) when calculated with Table Physics in contrast with Model Physics.

The source neutrons have equally probable energies of 1, 15, 100, and 150 MeV, for an average of 66.5 MeV. The secondary particle production differs significantly between the table and model physics.

The secondary angle distributions also differ between the table and model physics, as shown in Figs. 5.4, 5.5, 5.6, and 5.7. The secondary energy distributions are shown in Figs. 5.8, 5.9, 5.10, and 5.11. For each table, the MCPLOT tally plotter command is provided as the second plot title.

Table 5.1 Example 5.2 summary table output for various particle types showing the number of secondary particles produced per weight = 1.0 source neutron. The results are from separate runs using Table Physics and Model Physics

	Table physics		Model physics	
	Particles	Energy	Particles	Energy
Neutrons	1.0000E+00	6.6454E+01	1.0000E+00	6.6454E+01
Protons	1.6420E−01	1.0811E+01	3.5272E−01	1.5349E+01
Deuterons	6.3970E−02	2.0045E+00	3.1318E−02	4.4241E−01
Tritons	9.8542E−03	1.0999E−01	8.0357E−03	1.0768E−01
Helions	0.0000E+00	0.0000E+00	8.4313E−03	1.0224E−01
Alphas	1.8417E−01	1.4416E+00	4.7261E−02	4.5435E−01

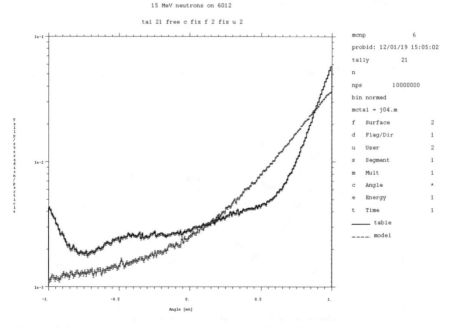

Fig. 5.4 Secondary neutron angular distribution of 15 MeV neutrons on ^{12}C

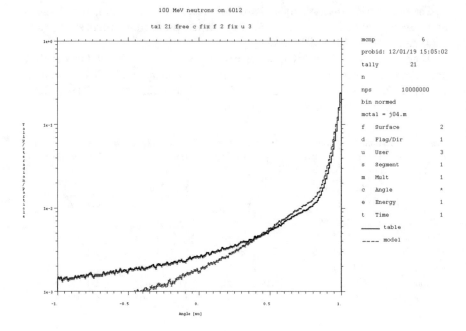

Fig. 5.5 Secondary neutron angular distribution of 100 MeV neutrons on ^{12}C

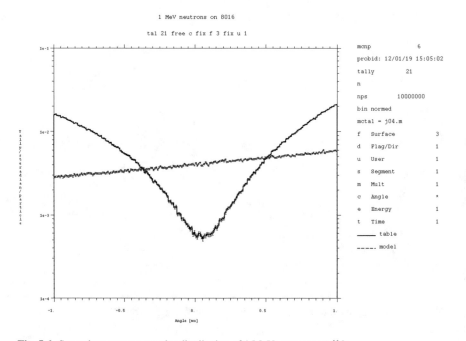

Fig. 5.6 Secondary neutron angular distribution of 1 MeV neutrons on ^{16}O

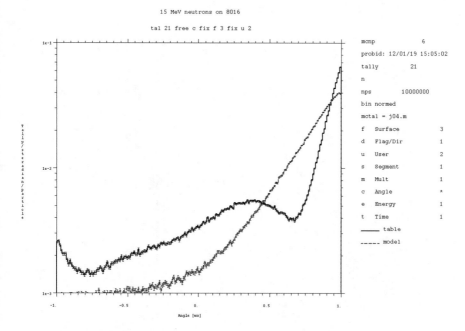

Fig. 5.7   Secondary neutron angular distribution of 15 MeV neutrons on ^{16}O

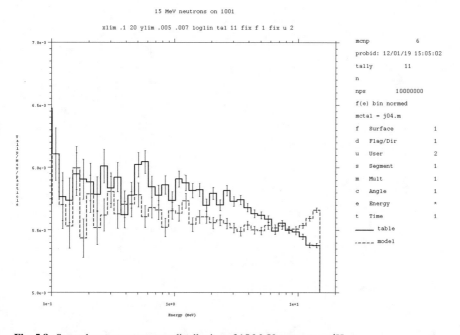

Fig. 5.8   Secondary neutron energy distribution of 15 MeV neutrons on ^{1}H

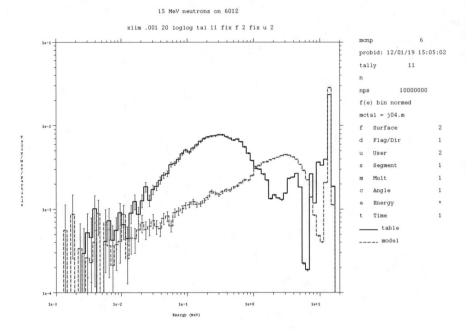

Fig. 5.9   Secondary neutron energy distribution of 15 MeV neutrons on ^{12}C

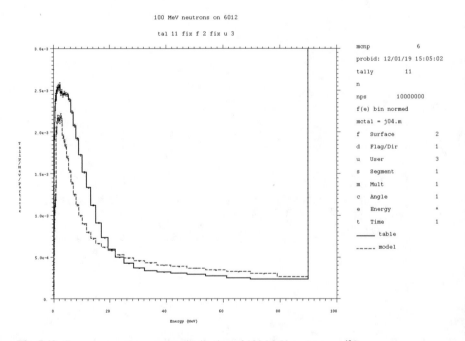

Fig. 5.10   Secondary neutron energy distribution of 100 MeV neutrons on ^{12}C

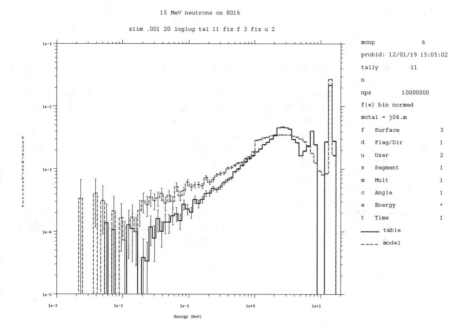

Fig. 5.11 Secondary neutron energy distribution of 15 MeV neutrons on ^{16}O

Figures 5.4, 5.5, 5.6, 5.7, 5.8, 5.9, 5.10, and 5.11 demonstrate that secondary particle production, energy spectra, and angular distributions can be estimated for single collisions of source particles with any target nuclei. In addition, the collision physics of table physics using nuclear data can be compared to model MCNP physics. Multiple comparisons can be made in a single run by having the different target nuclei in separate independent cells and by separating different incident energies with the FT SCX tally capability. Comparisons also could have been made between different physics models. Here, the default CEM model physics was used, but the ninth and tenth LCA entries could be used to select Bertini or other physics models. Other physics options such as EFAC (14th entry on the PHYS card) and HSTEP (on the M materials card) can also be compared (MCNP manual [1]) in Sections 3.3.3.2 and 3.3.2.1, Table 3-32, respectively.

References

1. P. Chard et al., A good practice guide for the use of modelling codes in non-destructive assay of nuclear materials. ESARDA Bull. **42** (2009)
2. R.J. McConn Jr., C.J. Gesh, R.T. Pagh, R.A. Rucker, R.G. Williams III, *Compendium of Material Composition Data for Radiation Transport Modeling, Revision 1*, Pacific Northwest National Laboratory report PNNL-15870 Rev. 1 (2011)

3. M.T. Swinhoe, C. Mattoon, S. Croft, I. Gauld, A. Nicholson, V. Mozin, *Nuclear Data Uncertainty Quantification – A Practical Example for Nuclear Measurements*, ANS Workshop, Santa Fe, NM (2016)
4. R. Weinmann-Smith, D.H. Beddingfield, A. Enqvist, M.T. Swinhoe, Variations in AmLi source spectra and their estimation utilizing the 5 Ring Multiplicity Counter. Nucl. Instrum. Methods Phys. Res., Sect. A **856**(C), 17–25 (2017)
5. A. Favalli, D.H. Broughton, S. Croft, M.S. Grund, R.D. McElroy, G. Renha, *Strengthening Technical Safeguards of Fresh Fuel through International Cooperation*, Los Alamos National Laboratory report LA-UR-19-31408 (2019)
6. D.P. Broughton, M.S. Grund, G. Renha, S. Croft, A. Favalli, Sensitivity of the active neutron coincidence collar response during simulated and experimental fresh fuel assay. Nucl. Instrum. Methods Phys. Res., Sect. A **1001**, 165243 (2021)
7. J.L. Conlin, *Listing of Available ACE Data Tables*, Los Alamos National Laboratory report LA-UR-17-20709 (2017)
8. GitHub (n.d.). https://github.com/NuclearData/DataListing
9. R. Weinmann-Smith, M.T. Swinhoe, A. Trahan, M.T. Andrews, H.O. Menlove, A. Enqvist, A comparison of Monte Carlo fission models for safeguards neutron coincidence counters. Nucl. Instrum. Methods Phys. Res., Sect. A **903**(C), 99–108 (2018)

Appendix: How to Get MCNP Software

MCNP software is distributed by the Radiation Safety Information Computational Center (RSICC) website at https://rsicc.ornl.gov/default.aspx.

Much useful information can be obtained from the Los Alamos National Laboratory MCNP website at https://mcnp.lanl.gov/index.shtml.

© The Editor(s) (if applicable) and The Author(s) 2022   293
J. S. Hendricks et al., *Monte Carlo N-Particle Simulations for Nuclear Detection and Safeguards*, https://doi.org/10.1007/978-3-031-04129-7

List of Figures

© The Editor(s) (if applicable) and The Author(s) 2022
J. S. Hendricks et al., *Monte Carlo N-Particle Simulations for Nuclear Detection and Safeguards*, https://doi.org/10.1007/978-3-031-04129-7

List of Tables

Index

Printed in the United States
by Baker & Taylor Publisher Services